T0359888

**MUSEUMS**VICTORIA
PUBLISHING

# DRAGON LIZARDS OF AUSTRALIA

## EVOLUTION, ECOLOGY AND A COMPREHENSIVE FIELD GUIDE

Jane Melville and Steve K. Wilson

Published by Museums Victoria Publishing


Museums Victoria Publishing
GPO Box 666
Melbourne VIC 3001
Australia
Telephone: +61 3 8341 7536
Web site: museumsvictoria.com.au/books

Design: ry • design
Printed by Focus Print Group, Melbourne

National Library of Australia
Cataloguing-in-Publication data

Dragon Lizards of Australia:
Evolution, Ecology and a Comprehensive Field Guide

ISBN 978-1-921833-49-6

A catalogue record for this book is available from the National Library of Australia

previous page:
Mallee Tree Dragon. *Amphibolurus norrisi*.
Border Village, SA
photo S. K. Wilson

right:
Chameleon Dragon. *Chelosania brunnea*.
Dampier Peninsula, WA
photo S. K. Wilson

contents page:
McKenzie's Dragon. *Ctenophorus mckenziei*.
Noondoonia Station, WA
photo R. Browne-Cooper

back cover:
Nobbi Dragon. *Diporiphora nobbi*. Male.
Girraween National Park, Qld
photo S. K. Wilson

# CONTENTS

## EVOLUTION, ECOLOGY & BIOLOGY



## GLOSSARY & QUICK GUIDE TO GENERA



## FIELD GUIDE

# EVOLUTION ECOLOGY & BIOLOGY

DRAGON LIZARDS OF AUSTRALIA

# INTRODUCTION

Australia is a land of lizards, with an amazing diversity of species that rivals any other country on Earth. Our mostly hot and dry continent is home to nearly 800 described lizard species. They belong to seven Australian families, which represent four evolutionary lineages: three families of geckos and flap-footed (legless) lizards; skinks; goannas; and dragons.

When many people think of a lizard they conjure an image of a dragon—perhaps a Frill-necked Lizard flaring its warning frill as it once appeared on our two cent coin, or a Boyd's Forest Dragon, the icon of Queensland's rainforests, or the prickly strangeness of a Thorny Devil from Australia's red centre, or even their pet bearded dragon. Australia's dragon lizards, currently numbering 102 species, show a remarkable variety of colour, form, ornamentation, ecology and behaviour.

The impressive spines and crests that adorn many dragon lizards may bring to mind the mythical dragons of story and legend, battled by many a brave knight. Though real dragons do not breathe fire and hoard jewels, they are every bit as intriguing and mysterious.

In this book we hope to shed some light on Australia's dragons. We will take you through the most recent understanding of their origins, life history, habitat and distribution. We have included an individual account of all of Australia's dragon species, featuring the most up-to-date taxonomic classification. Each is described, illustrated and mapped. We have great admiration for these amazing animals and are pleased to share the knowledge and experience we have gained from working with them. We hope that you enjoy reading this book as much as we did writing it, and that it will become a valuable reference and useful identification guide to Australia's amazing dragon lizards.

Ctenophorus cristatus.

Karalee Rock, WA.

Photo S. K. Wilson

# WHAT IS A DRAGON LIZARD?

Dragons belong to the family Agamidae, part of an ancient worldwide group called the iguanians (Infraorder Iguania). Based on fossil evidence, this ancient group—Iguania—has been around for more than 80 million years, and roamed the Earth alongside the dinosaurs (Evans & Jones 2010). Included in the Iguania are three lizard lineages, the dragons (Family Agamidae), the iguanas (Superfamily Iguanoidea, which includes a number of separate iguana families) and the chameleons (Family Chamaeleonidae). Of these three groups, the Iguanoidea occur mainly in the New World (North and South America), but also the Fijian Islands and Madagascar. The other two families, chameleons and dragons, have Old World distributions. Chameleons, which are the nearest living relatives of dragons, are found in Madagascar, Africa, the Middle East, Southern Europe, and east through India to Sri Lanka. Dragons are widely distributed from Africa, Southern Europe, the Middle East, through Asia to Australia.

An important distinguishing feature of dragons is their dentition. They have primarily acrodont teeth. Rather than sitting in sockets like our teeth, they are fused to the bone on the upper edge of their jaws, effectively forming serrated edges. They share this feature with chameleons. Acrodont teeth are not replaced. Dragons also have some pleurodont teeth at the front of the mouth. These are teeth that sit in sockets on a shelf on the inner side of the jaw, and individual teeth can be replaced. Iguanas (Superfamily Iguanoidea), along with most other lizard families, including all non-dragon lizards in Australia, have only pleurodont teeth. The two types of teeth are particularly obvious on a Frill-necked Lizard, with smaller acrodont teeth and canine-like pleurodont teeth, which are impressively large for a lizard.

There are other significant features that characterise dragons. All species have well-developed limbs, each with five clawed fingers and toes. Their tails, usually long and slender, lack the specialised weakened cleavage points of geckos, flap-footed lizards and most skinks. This means dragons' tails

**WHAT IS A DRAGON LIZARD?**

The Infraorder Iguania includes three broad groups of lizards—chameleons, iguanas and dragons. Chameleons are a highly distinctive family with independently mobile eyes, prehensile tails and grasping, mitten-like fingers and toes. Dragons and iguanas are much more similar to each other, distinguished mainly by dentition and distribution.

opposite page:
A gaping Mulga Dragon (*Diporiphora amphiboluroides*) exposes its enlarged, pointed pleurodont teeth. Such teeth surely mean business, and it has been suggested that they may play a role in interactions between males, or perhaps to aid grasping the female during mating. Munjina area, WA
photo B. Bush

## CHAMELEONS

## IGUANAS

## DRAGON LIZARDS

can be broken, but not shed (autotomised) as they can in these other families. If a dragon does lose its tail, it has a limited capacity for regeneration, producing little more than a stump rather than a new tail.

Dragons also share an alert, upright posture and loose skin with non-glossy scales. They are strongly visually-cued to locate prey, evade predators and engage in social behaviour. And like the iguanas and chameleons, they exhibit an enormous range of colour, both cryptic and for display. In many species, the males are more brightly coloured than females, particularly during the breeding season, and some can change colour. Many dragons also have some form of ornamentation, including spectacular arrays of thorn-like spines, raised crests, dewlaps, frills and beards.

All dragons are diurnal, which means they are active during the day. In Australia, they are all oviparous, and lay their eggs in purpose-built burrows rather than giving birth to live young. Their clutches of leathery-shelled eggs can number from two or three to 25 or more.

Most Australian dragons eat arthropods, typically insects and spiders. Some also consume small amounts of vegetation and this is particularly true for larger species, which follow a general trend in shifting towards an omnivorous or herbivorous diet. Most are sit-and-wait foragers, who keenly watch for moving prey from a vantage point before dashing over to grab the tasty morsel. However, some dragons actively hunt for prey and others regularly glean vegetation for edible flowers, shoots and fruits.

Australian dragons are currently divided into 15 genera. They are found in all states, though the Mountain Dragon (*Rankinia diemensis*), common in sub-alpine regions on the mainland, is the only dragon present in Tasmania. And they are found across most environments—from cool temperate heathland to the deserts of the interior and the rainforests of the wet tropics. Arid zones and the northern tropical woodlands support the greatest diversity of dragons. In these areas, they are probably the most commonly seen lizards. Their preference for sitting on elevated perches, such as fence posts, rocks and logs, ensures their visibility, while their complex behavioural repertoire—which includes hand-waves, tail-flicks and head-bobs—makes them worthy subjects for observation.

opposite page: (top to bottom):

Family Chameleonidae: Rwenzori Bearded Chameleon (*Triocerus rudis*). Besoke, Rwanda

photo S. K. Wilson

Superfamily Iguanoidea: San Cristobal Larva Lizard (*Microlophus bivittatus*). San Cristobal, Galapagos Islands, Ecuador

photo S. K. Wilson

Family Agamidae: Eastern Bearded Dragon (*Pogona barbata*). Glenmorgan, Qld

photo S. K. Wilson

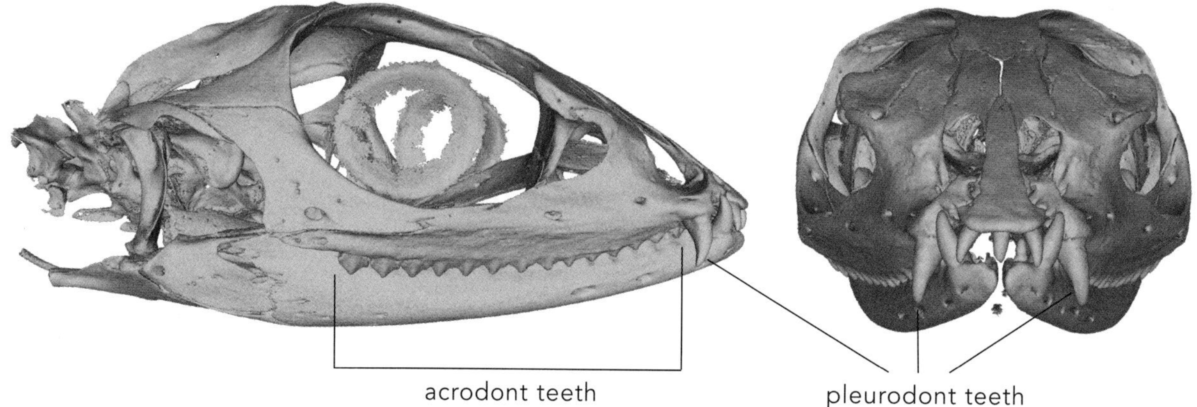


These CT scans of a male Yellow-sided Two-lined Dragon (*Diporiphora magna*) skull viewed from the side and the front, clearly show the acrodont teeth, arranged as fused serrations along the jaw-line and the more pointed pleurodont teeth located at the front of the jaw. The pleurodont teeth appear fearsome and fang-like. Yet the inoffensive little lizard, with a snout-vent length of just 68 mm, feeds only on small invertebrates.

photo J. Melville

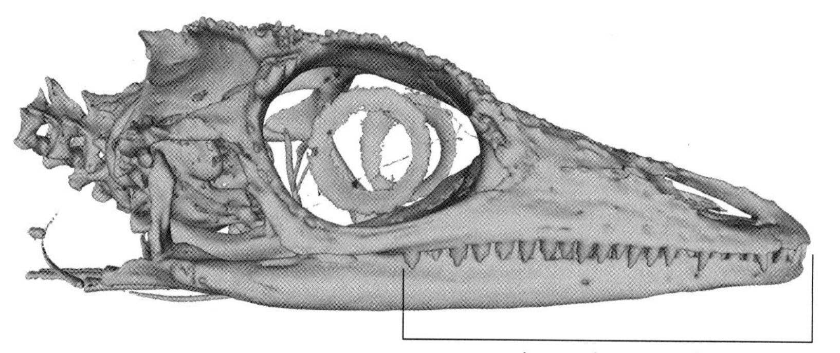

The jaw of an iguana features all pleurodont teeth, seen on this La Palma Anole (*Anolis insolitus*) skull viewed from the side and the front. Iguanas do not have any acrodont teeth.

photo E. Sherratt

# ORIGINS AND DIVERSIFICATION

Agamidae is classified into six sub-families, with all the dragons in Australia and neighbouring Papua New Guinea forming one of these lineages—the sub-family Amphibolurinae (Pyron *et al*, 2013; Zheng & Wiens 2016). The Chinese Water Dragon (*Physignathus cocincinus*) of Indochina also belongs to the Amphibolurinae and is the closest Asian relative to all Australasian dragons. The similarity between the Chinese Water Dragon and the Australian Water Dragon (*Intellagama lesueurii*) is striking, and until quite recently they were classified together under the one genus, *Physignathus*.

Based on recent genetic research, it is believed that Australasian dragons of the sub-family Amphibolurinae originate from a single migration event from South-East Asia 25–34 million years ago (Hugall *et al*, 2008; Melville *et al*, 2011). This was at the start of the Oligocene period (about 23 to 34 million years ago), when Australia was further south than it is today, abutting Antarctica, and the world was a warmer and damper place. Australia, covered with dense, lush rainforest, was just beginning its journey northwards toward Asia. This northward movement started an epic period in Australia of gradual continental climate change and opened up faunal migration from the north.

As well as the dragon lizards, lineages of geckos, skinks and goannas were among the new arrivals during the Oligocene (Oliver & Hugall 2017). The ancestral dragon probably looked something like the water dragons today. The subfamily has since diversified into the large radiation of dragon lizards that now exist in Australia, Papua New Guinea and a few of the southerly Indonesian Islands.

The early Australian dragons were probably forest inhabitants, with evidence from Riversleigh in western Queensland suggesting that ancestors of the genera *Lophosaurus* and *Intellagama* inhabited the tall wet forests of the Oligocene.

Genetic work shows that most of the basal genera of Australasian dragon lizards (meaning those with much older origins than other genera) are wet forest species, including *Hypsilurus*, *Lophosaurus* and *Intellagama*. Two exceptions to this are the monotypic genera *Chelosania* and *Moloch*, which are basal genera that occupy dry-to-arid environments (Schulte *et al*, 2003; Hugall *et al*, 2008). These ancient genera of the Oligocene were probably much more diverse. However, today they are represented by a few descendant species that have managed to survive in moist enclaves, remnants of forested habitats which used to cover much more of the continent. More work is needed on Oligocene fossil sites in Australia to understand more fully what our lizard diversity looked like.

Most of the Australian agamid genera are younger, with genetic evidence suggesting their origins lie in a burst of diversification during the Miocene period, about 10–20 million years ago (Hugall et al, 2008; Melville *et al*, 2011). This was the beginning of a drier continental Australia (Byrne *et al*, 2008).

The subfamily Amphibolurinae occurs throughout the Australopapuan region with an outlying species in Indochina. Horner's Dragon (*Lophognathus horneri*). Kununurra, WA
Photo: S. K. Wilson

With the final break-up of the great southern continent of Gondwana, South America, Australia and India moved north and global ocean currents changed. Instead of warm tropical waters moving to far southern latitudes, a circumpolar current began cooling the Earth. These continental changes altered Australia's climate; the wet forests retreated and the first signs of an arid environment started in the mid-Miocene, around 15 million years ago. The origins of the genera *Ctenophorus*, *Diporiphora*, *Pogona*, *Lophognathus*, *Chlamydosaurus*, *Gowidon*, *Tropicagama* and *Tympanocryptis* date back to this period and they all form a single evolutionary lineage that diversified in the new drier environment.

As aridification continued through the end of the Miocene and into the Pliocene, 2–10 million years ago, there was a huge burst of diversification in Australian dragons. Some clusters of related species (generally referred to as species groups) evolved around this time. For example, genetic evidence suggests that the expansion of grasslands in eastern Australia during the Pliocene, 2–5 million years ago gave rise to the grassland earless dragons (*Tympanocryptis pinguicolla*, *T. condaminensis* and their relatives; Melville *et al*, 2007; 2014).

Members of the arid-adapted *Ctenophorus maculatus* species group, often known as sand or military dragons, appear to predate the origins of the environments they now inhabit: the sand deserts (Edwards *et al*, 2015). These deserts are relatively young, not more than 1 million years old, so it has been hypothesised that the *C. maculatus* group may have originated in coastal sand environments and expanded inland as the sand deserts arose. Such a theory remains speculative but provides a scenario that explains patterns seen in genetic data.

Fossil evidence shows other twists and turns in the evolution and dispersal of Australian dragons, including more recent Pleistocene (11,700 to 2.6 million years ago) deposits at Mt Etna in Queensland. They suggest changes in the dragon lizard communities during a climate oscillation from cold glacial periods to warm interglacial periods. Species with moister requirements such as *Lophosaurus* and *Intellagama* were favoured in warmer, wetter interglacials at the expense of more arid-adapted species, such as *Pogona* and *Tympanocryptis*, which had prospered during cold, dry glacial periods (Hocknull *et al*, 2007).

Much remains to be learned about the origins of Australian dragons, and what we do know will certainly be refined with new information. However, what is certain is that dragon lizards are indelibly linked to their environment, and the dramatic climate changes that Australia has undergone over the last 34 million years has been a fundamental factor in shaping their evolution, diversity and distributions.

Agamid subfamily Agaminae extends through Africa and the Middle East to Europe and northern India and Pakistan. Secret Toad-headed Agama. (*Phrynocephalus mystaceus*). Ayaguz River, eastern Kazakhstan
Photo: M. Pestov

left:

Agamid subfamily Draconinae occupies Asia and South East Asia. Rhino Horn Lizard (*Ceratophora stoddartii*). Nuwara Eliya, Sri Lanka

Photo: S. K. Wilson

top:

Agamid subfamily Hydrosaurinae is restricted to eastern Indonesia and the Philippines. Sailfin Dragon (*Hydrosaurus amboinensis*). Camba, Sulawesi, Indonesia

Photo: D. Knowles

bottom:

Agamid subfamily Leiolepinae occurs in Malaysia and Indochina. Butterfly Lizard (*Leiolepis belliana*). Berjaya, Pulau Langkawi, Malaysia

Photo: G. Stephenson

top left:

Agamid subfamily Uromastycinae extends from the Middle East to Pakistan and northern India. *Uromastyx aegyptia*. Saudi Arabia

Photo: P. Wagner

top right:

The Chinese Water Dragon (*Physignathus cocincinus*) is more closely related to Australian dragons than to other South East Asian species. Lizards like this may have given rise to the Australian dragon radiation. Me Linh Station for Biodiversity, North Vietnam

Photo: D. Knowles

left:

During the Oligocene period, Australia had extensive moist forests. Ancestors of today's rainforest dragons (*Lophosaurus*) were much more widespread and diverse during this period than they are today. Southern Angle-headed Dragon (*Lophosaurus spinipes*). Mt Glorious, Qld

Photo: S. K. Wilson

top:
The expansion of eastern Australian grasslands 2-5 million years ago heralded the diversification of earless dragons. The four grassland earless dragon species are confined to fragmented temperate grasslands west of Melbourne, in ACT and in small areas of NSW. Monaro Grassland Earless Dragon (*Tympanocryptis osbornei*), NSW

Photo: S. K. Wilson

bottom:
Sand and Military Dragons (*Ctenophorus maculatus* and its relatives) appear to be older than the sandy deserts they now occupy. As the continent became more arid their ancestors may have moved inland from sandy coastal environments. Central Military Dragon (*Ctenophorus isolepis*). Alice Springs area, NT

Photo: R. Glor

# CONVERGENT EVOLUTION

Dragons and iguanas are clearly related lineages, though they occupy mutually exclusive global distributions. Aside from their dentition, there is little to distinguish them externally. They are both visually-cued lizards with upright postures, non-glossy scales, loose skin and a tendency for ornamentation with spines, tubercles or enlarged scales.

When faced with similar environmental constraints and opportunities there is a trend for both groups to follow the same morphological and behavioural trajectories. The resulting resemblances sometimes far exceed a passing similarity. In fact, the likeness can be so striking that some dragons and iguanids resemble each much more closely than they do members of their respective families. On opposite sides of the world, in the parallel universes of rainforests and deserts, in trees, on rocks or on open ground, these separate lineages of lizards have become each other's ecological analogues. This is convergent evolution.

The convergent evolution of desert dragons in Australia and iguanas in the south-western deserts of the United States is significant (Melville *et al*, 2006). In the arid stony regions of Western Australia, some of the earless dragons (*Tympanocryptis pseudosephos* and related species) closely mimic the gibber pebbles that surround them. Their heads and bodies are rounded like two stones and their coloration is a perfect match for the substrate. In the stony deserts of the south-western United States and northern Mexico, the Round-tailed Horned Lizard (*Phrynosoma modestum*) similarly reflects its landscape. Similar evolutionary convergence can be seen between independent lineages of dragons found in the Australian deserts and those in the deserts of central Asia. The Sunwatcher Agamid (*Phrynocephalus helioscopus*) from the central Asian stone-steppe deserts closely resembles the Australian earless dragons and North American horned lizards. These lizards crouch and tuck their short limbs close to the body if approached. There, among the stones, they become invisible. They have taken camouflage among reptiles to a rare level of sophistication. Rather than merely using colour and texture to blend in with their backgrounds, they actively mimic the inanimate objects around them.

With its dumpy body, short limbs, and armoury of thorn-like spines, the Thorny Devil (*Moloch horridus*) looks absolutely unique. True, nothing closely resembles this inhabitant of Australia's sandy deserts, but some of the larger horned lizards (*Phrynosoma cornutum* and other species) of North America and Mexico share some striking attributes. They feed exclusively on ants and, accordingly, have small mouths suitable for dabbing prey rapidly with their short tongues, reduced teeth (although their tooth specialisations differ; Meyers & Herrel 2005), large stomachs to allow extended feeding along ant trails, cryptic colouration, slow movement and spiny bodies. They also share the unique habit of 'rain harvesting', where moisture that contacts the skin is diverted via microscopic channels to the corners of the mouth, where it is consumed.

Rainforests pose different challenges for lizards, notably diffuse or scattered light in a complex world of shade and shapes. In Australian tropical and subtropical rainforests, the rainforest dragons (*Lophosaurus boydii* and *L. spinipes*) are slow-moving arboreal lizards that cling motionless to trunks and saplings in muted forest light. They frequently remain on the same perch for hours, where their angular heads, laterally compressed bodies and spiny dorsal crests help disrupt their outlines making them difficult to see in the shadowy forest. In the rainforests of South America a number of iguanids (including *Enyalioides* and *Corytophanes* species) can be found engaging in similar behaviours. Like its Australian counterparts the Amazon Forest Dragon (*Enyalioides laticeps*) has an angular head and spiny dorsal crest, tends to slide discreetly from view rather than dash for cover, spends hours motionless on a vine or slender stem, and sleeps on these sites at night.

The many examples of convergence between dragons and iguanas show how evolution repeatedly shapes species that live in separate but similar environments into forms that are strikingly alike. Yet, there are also some Australian dragon lizards for which there are no analogues. No parallels to the unique Frill-necked Lizard (*Chlamydosaurus kingii*) exist anywhere in the animal world. That extraordinary lizard really is a one-off!

With their compact spiny bodies, short limbs and tails, Australia's Thorny Devil (*Moloch horridus*) (right) and North America's Texas Horned Lizard (*Phrynosoma cornutum*) (left) have much in common. They even share small down-turned mouths, suitable for dealing with their specific prey–ants. Alice Springs area, NT; Roadforks, New Mexico, USA

Photos: S. K. Wilson

top:
The Goldfields Pebble-mimic Dragon (*Tympanocryptis pseudosephos*) is one of several species in Australia's western arid zones. Twin Peaks Station, WA

Photo: S. K. Wilson

bottom:
In central Asia, the Sunwatcher Agamid (*Phrynocephalus helioscopus*), also bears a close resemblance to pebbles. Panfilov District, eastern Kazakhstan

Photo: M. Pestov

A world away, in the deserts of New Mexico, the pebble-mimicking Round-tailed Horned Lizard (*Phrynosoma modestum*) belongs to the iguanian family Phrynosomatidae. Roadforks, New Mexico, USA
Photo: S. K. Wilson

It takes a keen eye to spot a pebble-mimic hiding in plain sight on stony ground! The art has been perfected by both dragons and iguanas in distant parts of the world.

The Southern Angle-headed Dragon (*Lophosaurus spinipes*) (left and above left) lives in subtropical eastern Australia. The Amazon Forest Dragon (*Enyaloides laticeps*) (above right and opposite page) is an iguana from South America. They occupy structurally similar rainforests on either side of the world and share very similar body shapes and life styles.
Photos: S. K. Wilson

top:
The long and the short of dragons. Those with short limbs, such as the Downs Bearded Dragon (*Pogona henrylawsoni*) are relatively slow-moving, tending to crouch or make a short dash for cover if disturbed. Winton area, Qld

Photo: S. K. Wilson

bottom:
The Crested Dragon (*Ctenophorus cristatus*) is a swift, long-limbed species that sprints on its hind limbs. Karalee Rock, WA

Photo: S. K. Wilson

# SIZE, SHAPE AND COLOUR

Australian dragons are a diverse group. They range in size from the Shark Bay Heath Dragon (*Ctenophorus butleri*) which could fit easily in the palm of your hand with an adult snout-vent length of 43 mm, up to the largest species, the Frill-necked Lizard (*Chlamydosaurus kingii*), where the snout-vent length of adult males may reach an impressive 300 mm.

There is a wide range of body shapes, including long-limbed and short-limbed species. Long-limbed dragons tend to be the swiftest. The sand and military dragons (*Ctenophorus maculatus* complex) are among the fastest lizards in the country. Other long-limbed species, including the Crested Dragon (*Ctenophorus cristatus*) and the semi-arboreal tree dragons (*Lophognathus*, *Gowidon*, *Tropicagama*) frequently run on their hind limbs (bipedal locomotion). Short-limbed species, including the bearded dragons (*Pogona*) and earless dragons (*Tympanocryptis*) tend to be slower and more reliant on camouflage. Those with the shortest appendages, the Gravel Dragon (*Cryptagama aurita*) and the Thorny Devil (*Moloch horridus*) are unique among Australian dragons in having tails shorter than their snout-vent lengths.

Differences in body shape are strongly indicative of lifestyle. Thanks to a laterally compressed body and tail, the semi-aquatic Water Dragon (*Intellagama lesueurii*) is a superb swimmer, while the extremely dorsally compressed head and body of the Ornate Dragon (*Ctenophorus ornatus*) allows easy access to the narrow crevices beneath granite exfoliations.

Although differences in size and shape can be quite dramatic, perhaps the most striking differences are the truly impressive arrays of colours, patterns and ornamentation. Over the last decade, Australian dragons have become important in understanding the evolution of lizard colouration and morphology.

The most obvious trend in colouration is for dragons to broadly match their backgrounds. Dragons occupying moist habitats of eastern Australia—the Water Dragon, Southern Angle-headed Dragon (*Lophosaurus spinipes*) and Boyd's Forest Dragon (*L. boydii*)—tend to have greenish hues. The Superb Dragon (*Diporiphora superba*), a foliage inhabitant of the northern Kimberley region, is bright green to match the leaves.

In the increasingly open arid habitats of the interior and west, overall body colouration is dominated by reds, yellows and browns. This is hardly surprising given that dragons are not only diurnal, they also tend to forage and perch in exposed sites to a much greater extent than other lizard families. Brick-red Long-tailed Sand Dragons (*Ctenophorus femoralis*) occupy exposed red sand dunes in Western Australia, while ring-tailed dragons of the *Ctenophorus caudicinctus* group match the reddish rocky ranges across much of arid Australia.

Some dragon species show marked variation in body colour between populations occupying different habitats. The Western Ring-tailed Dragon (*C. caudicinctus*) is typically red-brown, matching the red Pilbara rocks in the north-west of Western Australia, but just 50 kilometres away along the coastline of the Great Sandy Desert, a population with slate grey body-colour lives on pale rock outcrops. Genetic work has confirmed that these differences are colour variation within a species (Melville *et al*, 2016).

Perhaps the most striking example of variation between populations within a species can be seen in the Peninsula Dragon (*Ctenophorus fionni*) from the Eyre Peninsula in South Australia. It is a rock-inhabiting species, but the peninsula's outcrops are widely separated by unsuitable habitats of woodlands, mallees, heaths and, more recently, agricultural land. While the colours of the female dragons are relatively consistent between the rocky ranges at the north of the peninsula to the islands off the south, regionally distinct male colours exhibit an extraordinary degree of variation. At least six colour forms are identified, some of them quite stunning. A similar scenario exists with populations of Ornate Dragons (*C. ornatus*) on isolated granite outcrops in southern Western Australia.

Body pattern obviously plays a central role in camouflage. The Canegrass Dragon (*Diporiphora winneckei*), an inhabitant of spinifex and cane grass, has brown or yellowish vertical stripes on its belly that closely match the grass stalks on which it climbs. And across the genus *Diporiphora*, the dorsal pattern is usually 2 to 3 pale longitudinal stripes overlaying a series of short dark crossbands. A variation of this disruptive linear and transverse pattern also occurs in many earless dragons. It must be highly effective design for disruptive concealment, as a similar theme has evolved numerous times in iguanian lizards across all continents.

The Thorny Devil (*Moloch horridus*) takes disruptive colour and pattern to a whole new level. It employs a striking combination of sharply contrasting orange, yellow and black blotches and stripes, which serve to effectively break the animal's outline.

Often females and juveniles have more cryptic patterns than sexually mature males (Smith *et al*, 2011). This within-species variation can be seen across most Australian dragon lizards and appears to be related to an evolutionary trade-off between camouflage and reproductive success. For juveniles and females, it is more important to be well camouflaged to avoid being eaten, but for sexually mature males the pressure is on to look impressive, thus maximising their chances of outperforming their rivals and attracting a mate.

Male Red-barred Dragons (*Ctenophorus vadnappa*) appear to tread a fine line between successful coupling and predation. On the rocky outcrops of the Flinders Ranges in South Australia, brilliantly coloured males, with vibrant blue vertebral zones and black and crimson barred flanks, perch and display conspicuously. In doing so, they suffer higher predation rates from raptors than drabber males and also, presumably, the reddish-brown and dark mottled females. They may be trading away safe colours for conspicuous hues that promote better mating chances (Stuart-Fox *et al*, 2003).

top:
The semi-aquatic Water Dragon (*Intellagama lesueurii*) is laterally compressed. It can propel itself through water with consummate ease. Brisbane, Qld
Photo: S. K. Wilson

bottom:
Given the abundance of green foliage in Australian rainforests and along water courses, there are surprisingly few green reptiles. The Superb Dragon (*Diporiphora superba*) is an exception. It inhabits foliage beside gorges in the northern Kimberley region. Surveyor's Pool, WA
Photo: S. K. Wilson

opposite page:
A dorsally flattened body allows the Ornate Dragon (*Ctenophorus ornatus*) to hide under narrow rock crevices. Charles Darwin Reserve, WA

Photo: S. K. Wilson

above:
The rich reds of the outback are quintessentially Australian hues. Against this backdrop, many dragons have adopted matching colours. Slater's Ring-tailed Dragon (*Ctenophorus slateri*) has a vast distribution across rocky areas of the inland. Wolfe Creek Meteorite Crater, WA

Photo: R. Glor

However, the importance of camouflage from aerial predators in open habitats can still be seen in male lizards, with many species having camouflaged backs (dorsal surface) and their signalling colours on their lateral and ventral surfaces. Research has documented this evolutionary trade-off in the *Ctenophorus fordi* species group (Edwards *et al*, 2015), and demonstrated that Ornate Dragons (*C. ornatus*) are well camouflaged and inconspicuous from above but signal ultraviolet reflecting colours from their throats (LeBas & Marshall 2000), enabling them to communicate information such as sexual status to other dragons.

Ornamentation can also play a role in protection from predators and concealment, as well as sexual display and thermal management. Among those species for which ornamentation has a defensive role, the most famous example is the Frill-necked Lizard. The thin scaly ruff folded like a cape around its shoulders can be instantly erected to form a disc, 30 centimetres across. The apparent size of the lizard is suddenly increased and, with its gaping mouth, it creates an intimidating impression.

The spiny 'beards' of the bearded dragons (*Pogona* species) are used to the same effect, though the erectable appendage is a spiny pouch beneath the throat. The similarity of both groups has led to some confusion, with bearded dragons often referred to as 'frilly lizards' in some parts of Australia.

The impressive spines of the Thorny Devil render the otherwise harmless lizards unpalatable, bordering on inedible. It is unlikely that snakes could normally eat them, and goannas would also be deterred, though there is a record of a Thorny Devil in the gut of a Gould's Goanna (*Varanus gouldii*) (E. Pianka pers. comm.) Some raptors appear to have little difficulty in devouring the lizards and they discard the prickly skin (C. Dickman pers. comm), but there can be no doubt the thorns offer a high level of protection.

Like many other iguanians, dragons are able to change colour, sometimes rapidly. This has several purposes: it may reflect the lizards' mood when confronted with rivals, potential mates or predators, aid thermoregulation and assist concealment. Some also take on paler hues when inactive at night. Active dragons, particularly males, often exhibit bold hues of reds, yellows, black and white, yet when captured they may rapidly assume drab, less contrasting colours. Conversely, when confronted with a mate or rival their colours may brighten.

Dragons are often darker in the cool mornings, angling their bodies towards the sun to maximise heat absorption. Bearded dragons are masters of this ability and can even change colour differently across body regions, meaning they can absorb heat on their backs but not throat and upper chest (Smith et al, 2016a). Such colour differences between the upper and lower body surfaces provides advantages in temperature management and signalling to communicate. It has also been shown that colour change for thermoregulation is fundamentally linked to camouflage, with bearded dragons that live in different coloured habitats showing patterns of thermal colour change that optimise concealment (Cadena et al, 2017). An individual bearded dragon moves between perches to optimise both camouflage and thermal requirements (Smith et al, 2016b).

All this variation in body colour, patterning and ornamentation in Australian dragons is a complex game of trade-offs, prioritising in turn camouflage, thermal requirements and the social displays that underlie mate selection and reproduction. We have only scratched the surface of understanding these competing factors and there is much that is not yet known.

opposite page
right:
Pale stripes overlying a cross-hatching of dark bars is a common dorsal pattern among lizards spanning a number of families world-wide. In Australia it is shared by many of the two-lined dragons of the genus *Diporiphora*, including this Lally's Two-lined Dragon (*D. lalliae*). Barkly region, NT

Photo: S. K. Wilson

left:
Sometimes bold markings help break an animal's outline, while narrow stripes further disrupt the image. The pattern effectively conceals this Thorny Devil (*Moloch horridus*) among dappled shade.
Bullara Station, WA

Photo: S. K. Wilson

# DISPLAYS

It is hardly surprising that dragon lizards rely strongly on colour, posture, ritualised movements and an array of spectacular appendages to communicate with each other. Eyesight is their primary sense, and these behavioural antics that result from their high visual acuity make dragons fascinating animals to observe.

Colours that can be subdued or intensified according to hormones in breeding season, health, and aging are a common feature of many dragons, particularly members of the genus *Ctenophorus* (Olsson *et al*, 2013). Males can brighten and intensify their colours during breeding season, which plays an important role in a lizard displaying to rivals or mates.

South Australian rock dragons of the *C. decresii* species group exhibit some of the most spectacular combinations of colour and movement when rival males encounter one another (Gibbons 1979). The male Red-barred Dragons (*C. vadnappa*), adorned with their bright blue vertebral region, black and red barred flanks and a blue and yellow striped throat, align themselves sideways to their opponent. They erect a vertebral crest of skin, distend their colourful throat, rhythmically raise and lower their bodies, and coil their tails vertically over their backs. The related Tawny Dragon (*Ctenophorus decresii*) performs a similar display, with the tail coiled horizontally, while the Peninsular Dragon (*C. fionnii*) coils its tail obliquely or horizontally.

Many dragon species also signal with their heads and forearms. Some perform an elaborate series of head bobs and dips and raising and lowering of the arms (circumduction). Jacky Lizards (*Amphibolurus muricatus*) perform their stunts from an elevated perch such as a rock, log or stump. Their signals

Male rock dragons of the *Ctenophorus decresii* species group perform spectacular displays on outcrops in SA and far western NSW. They raise their bodies high on their legs, gape their mouths and distend their throats. Depending on species, they also coil their tails vertically, horizontally or obliquely. The Barrier Range Dragon (*C. mirrityana*) occurs in NSW. Mutawintji National Park
Photo: J. de Jong

opposite page:
With an erectable frill up to 30 centimetres across, the Frill-necked Lizard (*Chlamydosaurus kingii*) has the proportionally largest display appendage in the reptile world. When the frill is erected and mouth agape it presents an extraordinary vista to rivals and potential predators. Arnhem Land, NT
Photo: R. Glor

have been found to comprise multiple distinct components, including tail-flicks, arm-waves, push-ups and body-rocks (Peters & Ord 2003). Under experimental conditions it was found that the resident lizard is much more likely to display than an intruder. It appears that the display serves the purpose of establishing dominance without the need for combat. 'The message gets across and no-one gets hurt.'

While displays are often perceived as the domain of males, the females of some species engage in eye-catching behaviour. Female Canberra Grassland Earless Dragons (*Tympanocryptis lineata*) from the ACT circle each other while performing exaggerated push-ups, with their bodies extremely laterally compressed, throats extended, legs fully stretched and tails horizontal. And males beware! Under captive conditions, they must be separated or the aggressive females will injure them (S. Sarre, pers. com). Meanwhile, on the exposed salinas in South Australia, female Lake Eyre Dragons (*Ctenophorus maculosus*) demonstrate an unwillingness to mate by threatening amorous males with their backs arched and bodies laterally compressed. As a last resort they will flip onto their backs to deter males, presenting a bright orange abdomen, to prevent mating.

Head-bobbing and arm-waving are widespread among dragons, but these actions are not always directed at mates or rivals. Various short sequences are often enacted when an animal is alone, and broadcast to no one in particular. Central Bearded Dragons (*Pogona vitticeps*) will perform a few head bobs or dips from a high vantage point without any obvious stimulus and no other dragon in sight. A canny human observer can even elicit further bobs by responding with a flick of the hand. And it is common for dragons of many species to wave their arms following a short burst of activity. A short sprint, a few waves, then another sprint. In Northern Australia, the Gilbert's Dragon (*Lophognathus gilberti*) and its relatives are so famous for their arm-waving that they are known colloquially as 'Ta-ta' or 'Bye-bye' Lizards.

Many dragons are well adorned with fancy 'extras' in the form of crests, spines and gular (throat) pouches (Ord & Stuart-Fox 2006). When displayed, with crests raised high and gular pouches extended, they create the illusion of increased size and often bear strongly contrasting colours and patterns. Because

opposite page:
Male Water Dragons (*Intellagama lesueurii*) are highly territorial, and aggressive displays and combat are more common in high-density urban populations. Here male Water Dragons in inner-city Brisbane can be seen grasping jaws (top). There are consequences to these encounters with most males in these high density populations exhibiting wounds to their snouts (bottom left) and this male appears to have a broken jaw (bottom right). Assuming it survives it will be scarred for life. Brisbane, Qld
Photos: S. K. Wilson

top:
Bearded Dragons (*Pogona* species) have an erectable spiny pouch on the throat. With the 'beard' extended, mouth opened and body flattened the lizard appears suddenly and dramatically larger. This male Eastern Bearded Dragon (*Pogona barbata*) also exhibits some jaw damage, a likely consequence of prior combat. Dutton Park, Qld.
Photo: S. K. Wilson

centre:
Some dragons have colours that can be selectively revealed. The intense throat colours on this male Peninsular Dragon (*Ctenophorus fionni*) are strategically placed so they can be visible to other dragons. By bobbing its head the colours can be flashed to convey information on territory and sexual status. Tumby Bay, SA.
Photo: S. K. Wilson

bottom:
The bright yellow in the corners of the mouth of this Western Heath Dragon (*Ctenophorus adelaidensis*) are only visible when the mouth is gaped. It is not known whether this targets potential predators or other dragons. Perth, WA.
Photo: S. Mahony

such displays are most effective when viewed laterally, lizards generally align themselves side by side, often enhancing their performance with other behaviours such as mouth gaping and body swaying. This posturing is common in territorial disputes and may also provide sexual cues during mate choice.

At about 30 centimetres across, the frill of the Frill-necked Lizard (*Chlamydosaurus kingii*) is one of the largest and most spectacular display structures proportionate to body size on a terrestrial vertebrate. The purpose of this scaly disc, almost completely encircling the neck, is to communicate with others of its kind and deter predators. Supported by slender hyoid bones (small bones in the throat that usually function to support the lizard's tongue) the frill can be erected as the mouth is gaped open.

Male Frill-necked Lizards display and fight during the mating season around November and December (Shine 1990) and combat is common. From perches on tree trunks, or occasionally on the ground, they audibly slap their tails, tilt their heads up and down, partially open their frills and wave their arms. They will lunge at each other and interlock their jaws. During this time, most adult males exhibit recent injuries to the jaw, ranging from fresh blood to missing teeth and even broken mandibles. Interestingly, it is the males with the brightest coloured frills, rather than the largest, which have the greatest success in these dramatic encounters (Hamilton *et al*, 2013). For females, immature males and juveniles the frill seems exclusively a predator deterrent as they do not appear to display to other Frill-necked Lizards.

Male Water Dragons (*Intellagama lesueurii*) also frequently engage in combat during the breeding season. Their impressive disputes, which include posturing, chasing and fighting, are particularly prevalent in populations where high densities occur, such as in some of Brisbane's inner city parklands (Strickland *et. al.*, 2018). At these sites, virtually all mature males bear the scars of combat.

Bearded Dragons (*Pogona* spp.) also have a display structure supported by hyoid bones. The spiny pouch or 'beard' is rapidly erected while the pink or bright yellow mouth is gaped. They can also further increase their apparent size by flattening their bodies dorsally to form a broad spiny disc edged with slender spiny scales. In some species, these postural displays are often associated with a dramatic and rapid change in body colour, with the beard and belly becoming black, and the flanks and back a contrasting yellow-to-orange hue. When tilted obliquely towards a rival or potential predator, in tandem with the beard and open mouth, the effect is striking.

The head-bobbing, tail-slapping, arm-waving antics of Australian dragons, many of which add an array of brilliant colours to their communication repertoire, are visually spectacular. Of continuing interest are their colour displays in the ultraviolet spectrum, and the trade-off where brightly coloured males attract both the interest of potential mates and the increased attention of predators. Communication among the dragon lizards is a complex affair that will clearly be the focus of research for many years to come.

The Central Netted Dragon (*Ctenophorus nuchalis*) is a burrowing lizard found over vast tracts of the inland. The sloping burrows it creates at the bases of trees and shrubs, and in windrows beside tracks are prime subterranean real estate, critical assets for countless other animals. Snakes, lizards and small mammals utilise the insulated retreats. Quilpie area, Qld
Photo: S. K. Wilson

# DISTRIBUTION AND HABITAT USE

The general trend among Australian reptiles is for diversity to be low in the south and much higher in the north and arid centre (Powney *et al*, 2010). Dragons are no exception, with very few species in the southern cool temperate regions, large numbers across the northern tropics, and exceptional diversity in parts of the arid zone vegetated with spinifex grasses (*Triodia* spp). Although rainforests are hailed as among the Earth's most biologically diverse habitats, in Australia they score poorly for a sun-loving group such as dragons, with just two species of *Lophosaurus*. This contrasts with the South-East Asian rainforests, the likely origin of Australian dragons, which remain rich in species.

Australia's most southerly agamid, the Mountain Dragon (*Rankinia diemensis*), is the only species in Tasmania and parts of the Victorian highlands. Within those cool climatic zones, often subject to regular winter snow, these dragons are largely confined to open sunny habitats such as heathland and the edges of forests (Ng *et al*, 2014).

Conversely, in the Exmouth region of Western Australia, an area that includes the stony ranges of North West Cape, sand dunes and flats, at least 15 species of dragons are recorded (Storr & Hanlon 1980). A similar number occur in parts of the tropics, such as areas in the Kimberley region where savannah woodlands abut water courses and sandstone massifs vegetated with spinifex.

To coexist in such numbers, dragons must partition the available resources. A single site may include some rock-dwelling, terrestrial and arboreal species, and each generally occupies particular micro-habitats, selected on the basis of preferred structural and thermal attributes. Ecological differentiation among co-occurring species of *Ctenophorus* is evident in the kinds of habitats used for shelter, which have been categorised as burrows, rocks or shrubs/hummock grasses (Melville *et al*, 2001).

The following scenarios are enacted across the continent in a myriad of complex ways as related species live different lives within shared habitats.

In a dry southern woodland at Lake Hurlestone in Western Australia, at least three species of *Ctenophorus* live a mere few hundred metres apart; a Crested Dragon (*C. cristatus*), views the world from the elevated perch of a fallen log, a Claypan Dragon (*C. salinarum*) hugs the edge of the low shrubs where its sloping burrow is within easy access, and, on the exposed granite dome that rises from the flats, Ornate Dragons (*C. ornatus*) perch on rocks and shelter under exfoliations. On a loamy desert flat in the Barkly region of the Northern Territory, a Central Netted Dragon (*C. nuchalis*) has a fine view from the top of a termite mound, placing its faith in the knowledge that several burrows are close if needed, and a Central Military Dragon (*C. isolepis*) dashes about on the ground below, seeking neither elevated perches nor burrows, for it inhabits a world of hot ground and shady spinifex. And beside a Kimberley water course on the Mitchell Plateau, a Superb Dragon (*Diporiphora superba*) clings to slender stems amid the thick green foliage of a shrub, but a Robust Two-lined Dragon (*D. bennettii*) prefers to remain on the ground, perched on a piece of sandstone in the dappled shade (SW pers. obs.), while a Pale Two-pored Dragon (*D. pallida*) climbs through the spears of a spinifex tussock (JM pers. obs.).

In the sand dune country of the eastern Simpson Desert, *Ctenophorus* species are known to partition the habitat. In that area, Central Military Dragons inhabit the slopes and crests of dunes, while the Central Netted Dragons (*C. nuchalis*) prefer the inter-dune flats (Dickman *et al*, 1999). In the Exmouth area, a number of closely related species occur. The Rufus Sand Dragon (*C. rubens*) lives its short but fast life on the inter-dunes at Giralia Station, while the slopes and crests are inhabited by the Long-tailed Sand Dragon (*C. femoralis*) (Wilson & Knowles 1988).

Some dragons are so dependent on rock that any intervening terrain is an effective barrier to dispersal. For the Ornate Dragon of Western Australia, and the Peninsula and Tawny Dragons (*Ctenophorus fionni* and *C. decresii*) of South Australia, isolated outcrops are inhabited by populations with strikingly different looking males, a result of long isolation on their rocky landlocked islands. A study of the Tawny Dragon found that local environmental conditions (aridity and vegetation cover), combined with low levels of genetic exchange across intervening terrains, influence the colour differences between populations (McLean *et al*, 2015). These species are all obligate rock inhabitants, perching on raised stones and sheltering beneath them.

However, phylogenetic analyses of *Ctenophorus* suggest that the ancestral condition is to use burrows for shelter (Melville *et al*, 2001) and reliance on burrows is seen across multiple genera, such as *Tympanocryptis*, *Diporiphora*, *Lophognathus*, and *Amphibolurus* (JM pers. obs.). Most dig their own (often several) but at least one Australian dragon is dependent on other organisms to construct the required burrows. For the Canberra Grassland Earless Dragon (*Tympanocryptis lineata*), a habitat specialist restricted to the highly fragmented native temperate grasslands, burrows excavated by arthropods are a critical resource, with individuals having one or two home burrows (Stevens *et al*, 2010).

Whether perched on a fence post or at the entrance to its burrow, in a tropical woodland or a desert flat, the different sites Australian dragons choose to occupy and the habitats where they occur are more than whim and chance. They are the ongoing by-products of constant selective pressures on animals that strive to thrive in harsh and complex environments.

# TEMPERATURE MANAGEMENT

Like all reptiles, dragons source their body heat externally. Put simply, they raise their temperatures by selecting warm sites, lower them by shifting to cooler areas, and maintain stability by shuttling between them. This contrasts with mammals and birds, which generate body heat internally. Reptiles are often called 'cold-blooded' (and mammals and birds 'warm-blooded'), but this term is misleading, having no relevance to the preferred operational body temperatures of reptiles.

In many cases, dragons operate at temperatures higher than those of most mammals. In fact, if the core body temperature of a human was raised to levels preferred by some desert dragons, that person would find themselves in a medical emergency.

Being diurnal, dragons generally raise their body temperatures by basking in the sun and cool off by moving into the shade or under shelter. As well as this, they employ a sophisticated combination of strategic use of colour and posturing to maintain an optimal body temperature, often within quite narrow parameters.

Many dragons adopt dark, heat-absorbent colours when they first bask in the morning. They also assume a flattened posture with their broad dorsal surfaces exposed to the sun. As they reach their optimal temperature they become paler, and are then able to engage in normal daily activities.

The use of colour to change absorptance (the effectiveness of a surface in absorbing heat from the sun) has been measured in some dragons. Changes in skin reflectivity have been found to vary directly with body temperature in Jacky Lizards (*Amphibolurus muricatus*) and Central Bearded Dragons (*Pogona vitticeps*), with their skin colour changing to reflect away more sunlight when their body temperatures are high (Rice & Bradshaw 1980). For the Western Ring-tailed Dragon (*Ctenophorus caudicinctus*) the difference between their light and dark body colour phases equates to about a 2°C difference in body temperatures (Christian *et al*, 1996). However, not all Australian dragons can control their temperature in this manner, with the same study finding that the Frill-necked Lizards (*Chlamydosaurus kingii*) do not have a pronounced ability to change absorptance with respect to body temperature. This means that some but not all dragons can thermoregulate by changing colour.

To warm up dragons bask but the poses they adopt to reduce heat-gain include a more varied and impressive repertoire. They angle themselves directly towards the sun to reduce the exposed surface area and they raise their bodies as high as possible from the ground, minimising contact with the hot substrate to reduce radiant heat from below. Some can remain on their perches during extremely hot weather by 'stilting', with body held aloft and only the claws of the front feet and the heels of the hind feet touching the surface.

A
B
22.0°C
17.5°C
C
D
42.0°C
19.0°C

opposite page:

This thermal imagery illustrates two critical elements in the daily thermoregulation of a Central Bearded Dragon (*Pogona vitticeps*). Early in the morning the lizard adopts dark colours to warm itself as quickly as possible (A), while its temperature is shown to be at the low end of the thermal scale (B). Later in the day it has selected pale reflective colours to avoid excessive heat gain (C), and it has achieved a more optimal temperature (D)

Photos: D. Stuart-Fox

this page

top left:

With its body flattened to the sun, a female Red-barred Dragon (*Ctenophorus vadnappa*) draws in the morning heat quickly and efficiently. Until it reaches optimal temperature it is less able to defend itself, flee danger and engage in other daily activities important for survival. Blinman Creek, Flinders Ranges, SA

Photo: S. K. Wilson

top right:

During the heat of the day, Eyrean Earless Dragons (*Tympanocryptis tetraporophora*) adopt an exaggerated upright pose. Like miniature dinosaurs they stand with body erect, supported by a tripod of hindlimbs and tail, angling themselves into the sun to minimise direct exposure. Hughendon-Winton Rd, Qld

Photo: S. K. Wilson

bottom:

This Centralian Earless Dragon (*Tympanocryptis centralis*) is beating the heat by climbing into a bush to elevate itself from the hot ground. West MacDonnell Ranges, NT

Photo: J. Melville

The Eyrean Earless Dragon (*Tympanocryptis tetraporophora*) thrives in the harshest, most featureless terrain on the continent: stony deserts and open cracking clay with very sparse vegetation. By careful posturing they are often, quite literally, the last animals standing in temperatures of 40°C and above. Standing only on their hind legs, usually on a small stone or earth clod, they prop themselves vertically like tiny bipedal dinosaurs, angled directly into the sun while their white undersides help deflect the heat radiating from the shimmering substrate.

Four thermoregulatory postures have been identified in the Central Military Dragon (*Ctenophorus isolepis*) (Losos 1987). In early morning, when temperatures are cool, they lie flattened to the ground. As the day warms they retreat to semi-shade with fore-body raised, and as the mercury rises the body is lifted high off the ground with only the tips of the feet and base of tail touching. During extreme heat, with ground temperatures above 50°C, all four legs are extended, lifting the body entirely off the sand, with tail rigid and held above the ground as well. All that is touching the ground are the heels of their feet and the curve of their tail. Variations on these postures have also been observed in Western Ring-tailed Dragons (*C. caudicinctus*), Central Netted Dragons, (*C. nuchalis*) and Ornate Dragons (*C. ornatus*) (Bradshaw & Main 1968) and the Yinnietharra Rock Dragon (*C. yinnietharra*) (SW pers. obs).

A study of nine species of central Australian dragons (Melville & Schulte 2001) found that thermoregulatory behaviour was important throughout a lizard's daily activity; all species, other than Central Military Dragons (*Ctenophorus isolepis*), were found to increase their perch height in the middle of the day. Central Military Dragons (*C. isolepis*) were shown to be strictly terrestrial and were never seen perching off the ground.

One desert species, the Canegrass Dragon (*Diporiphora winneckei*), is recorded to have survived nearly half an hour with a body temperature of 49°C (Bradshaw & Main 1968). It has been observed clinging to the top of vegetation at the crest of dunes in the Simpson Desert, presumably to lower body temperatures in the breeze, on days when ground temperatures were greater than 65°C (JM pers. obs). During field observations, lizards perched on vegetation recorded body temperatures of 42°C–46°C while the air temperature was 45.5°C (Greer 1989).

The amount of time spent in the sun versus shade depends on the time of day, year and local climate. A temperate species such as the Mountain Dragon (*Rankinia diemensis*) is likely to spend more time basking while dragons occurring in desert and tropical regions allocate more time in the shade or even sheltering in burrows and under rocks during hot weather. When active, Ta-ta Lizards (*Lophognathus* spp.) are most often found in full shade and within 5 metres of vegetation cover (Thompson & Thompson 2001).

If they get too hot, some dragons use evaporative cooling by gaping their mouths. Central Bearded Dragons gape at an average temperature of 36.9°C (Tattersall & Gerlach 2005). This study also reported cloacal discharge at high temperatures, suggesting that evaporative water loss from the cloaca may also play a role in temperature regulation. Gaping has also been observed in Central Military Dragons and many other species.

top:
Some dragons can tolerate extraordinarily high temperatures. While the shade temperature is 42 degrees this juvenile Smooth-snouted Earless Dragon (*Tympanocryptis intima*) is subjecting itself to a much hotter scenario, perched on a dark metal chain in full sun. Winton area, Qld
Photo: S. K. Wilson

bottom:
Panting is another means of avoiding overheating. The Military Dragon (*Ctenophorus isolepis*) pictured is employing evaporative cooling by drawing air over the moist mouth-parts. Windorah, Qld
Photo: S. K. Wilson

Many dragons like this Yinnetharra Rock Dragon (*Ctenophorus yinnietharra*) raise their bodies, and minimise contact with the hot substrate by supporting themselves on the heels of the hindlimbs. Pale ventral surfaces deflect radiant heat from below. Yinnetharra Station, WA
Photo: S. K. Wilson

Rainforest dragons (*Lophosaurus* spp.) are sit-and-wait predators with a reliance on immobility for ambush and camouflage. It is not viable for them to chase errant puddles of sunlight within a rainforest so they have become thermoconformers, meaning they allow their body temperature to rise and fall with the ambient temperatures around them. Body temperatures of five radio-tracked Southern Angle-headed Dragons (*L. spinipes*) were low and variable (11°C–26°C). A minor but significant behavioural element that may have elevated the lizards' temperature was a preference for high vertical perches of intermediate diameter in areas with relatively open canopy (Rummery et al, 1995). These dragons, in their cool, shaded environment, are the least reliant on direct sunshine of all the Australian dragons.

# DIET

Dragons are alert, visually-cued predators. With an upright stance they survey their surrounds, often from an elevated vantage point ranging from a slightly raised stone or soil clod to a branch, fence post or tree trunk. While some, such as the sand and military dragons (*Ctenophorus maculatus* and its relatives), actively forage, many species adopt a sit-and-wait approach to hunting. Rainforest dragons (*Lophosaurus* spp.) spend hours clinging to a sapling or trunk, scanning the leaf litter below for telltale signs of insect activity.

Frill-necked Lizards (*Chlamydosaurus kingii*) behave in much the same way as they perch on rough-barked tree trunks a metre or two above the ground in the Top End tropical woodlands. On spying insects on the ground, they race from their vantage points to seize their prey, and, unlike most lizards, they commonly walk on their hindlegs (bipedal locomotion) during routine foraging (Shine & Lambeck 1989). Ta-ta Lizards (*Lophognathus* spp.) also hunt from an elevated perch, and it is estimated that they catch a prey item every 92 minutes, or 6-7 items each day (Thompson & Thompson 2001).

Prey is detected by movement, pursued and captured with a dab of the tongue, crunched with the teeth and swallowed. Employing the tongue to capture prey is a uniquely iguanian feature. In dragons and iguanas the tongue is short and thick and just slightly protruded to pick up prey. The most famous iguanian tongues are those of chameleons—they are high-speed projectiles that may exceed the lizards' own body length.

The finding that Australian dragon lizards, in particular the Eastern Bearded Dragon (*Pogona barbata*), have protein-secreting glands producing various enzymes and peptides along the upper and lower jaws raises the possibility that more than just mechanical action may be used to consume prey. It has been suggested that these glands are involved in prey capture or pre-digestion or both (Fry *et al*, 2006).

For the most part, dragons feed primarily on arthropods. By virtue of sheer abundance the bulk of their diet consists of insects. Spiders, other invertebrates, including centipedes, and smaller vertebrates, including other lizards, are also taken opportunistically.

Dragons eat a wide range of insects, provided they are suitably sized and non-toxic. They also regularly exploit an insect food resource that is often shunned by other lizards—ants. Many ants contain formic acid, which may be why other diurnal insectivorous lizards, such as skinks, tend to ignore them. Yet many small dragons rely on them as a primary food source. An examination of the gut contents of Central Military Dragons (*Ctenophorus isolepis*) from the sandy deserts revealed ants comprised

top right:
Dragons opportunistically eat vertebrates. This Central Bearded Dragon (*Pogona vitticeps*) is endeavouring to swallow an adult Central Netted Dragon (*Ctenophorus nuchalis*). Ballera, Qld
Photo: S. K. Wilson

top left:
Using the tongue to pick up food is a common and distinctive theme among iguanians worldwide. Chameleons' tongues are long, high speed projectiles but those of dragons and iguanas are short and thick. This Central Bearded Dragon (*Pogona vitticeps*) is capturing a lacewing
Photo: S. K. Wilson

centre and bottom:
A common theme across many lizard families is a shift towards an omnivorous or herbivorous diet among the larger members. This includes heavyweights such as Galapagos Land Iguanas (*Conolophus subcristatus*) (middle) which are almost exclusively vegetarian as adults, and Water Dragons (*Intellagma lesueurii*) (bottom) which enjoy a broad omnivorous diet. middle photo Santa Cruz, Galapagos; bottom photo Mt Coot-tha, Qld
Photo: S. K. Wilson

opposite page:
Most dragons feed mainly on invertebrates, particularly arthropods. This Water Dragon (*Intellagma lesueurii*) is making short work of a spider. Mt Coot-tha, Qld
Photo: S. K. Wilson

94 per cent of prey items (Pianka 1971a). In Western Australia's acacia shrublands, a study examining Lozenge-marked Dragons (*C. scutulatus*) found that most individuals (96%) had eaten ants (Pianka 1971b).

The Thorny Devil (*Moloch horridus*) is an ant specialist, consuming two common types of small ants (Withers & Dickman 2005): primarily *Iridomyrmex* species from trails on the ground but also *Crematogaster* sp. from trails along the stems of the native currant bush (*Leptomeria preissiana*). From this research it was estimated that Thorny Devils eat approximately 750 ants each day. The slow-moving lizard, walking like a jerky animated clockwork toy, positions itself above an ant trail. Every ant is a potential target as they file within striking range of that short sticky tongue. The Thorny Devils differ significantly from other dragons in some aspects of their feeding (probably to do with their specialist diet). They dab with their tongues but they do not lunge at their prey; they have a faster tongue protrusion and there is reduced processing of their food (Meyers & Herrel 2005). Curiously, an examination of Thorny Devil scats reveals little or no sand grains (SW pers. obs.). All those strikes on the swift little insects hurrying across sandy ground and every one a direct hit!

Sometimes much larger prey are taken. A Jacky Lizard (*Amphibolurus muricatus*) has been observed capturing small skinks and a Central Bearded Dragon (*Pogona centralis)* was observed trying to swallow a Central Netted Dragon (*Ctenophorus nuchalis*) that was so large the tail remained protruding even after the bulk of the lizard had been ingested (SW pers. obs.).

Members of the family Agamidae conform to a world-wide trend showing broad dietary differences with body size. The larger a species the more omnivorous or herbivorous it is likely to be (Cooper & Vitt 2002). This trend is also evident in other families, including iguanas, skinks and lacertids. In Australia, omnivorous dragons with a large body size include the bearded dragons (*Pogona* spp.) and the Water Dragon (*Intellagama lesueurii*).

However, the juveniles and sub-adults of these lizards, which are approximately the adult size of other smaller dragon species, are mainly insectivorous. They consume increasing amounts of vegetation as they grow; as adults this includes significant quantities of flowers, berries and soft new vegetation.

But there are significant exceptions. The Frill-necked Lizard Lizard is Australia's largest agamid and appears to be almost exclusively insectivorous, relying on large numbers of small items, such as caterpillars, Green Tree-Ants (*Oecophylla smaragdina*) and termites. One individual Frill-necked Lizard contained about 1250 termite alates. If taken in one sitting, these smaller items offer equivalent bulk to a single larger prey item. The minimal amount of plant material recorded (in only two of 124 stomachs) was probably ingested accidentally. Vertebrates were also rare items, with a juvenile death adder perhaps mistaken for a caterpillar (Shine & Lambeck 1989), and rare records of birds and frogs being taken (McKay 2011).

# REPRODUCTION

Virtually all members of the family Agamidae across the world are egg-layers. Just two Sri Lankan dragons, *Cophotis* species, from high altitude mossy forests, and five *Phrynocephalus* species from the high Qinghai-Tibet Plateau in western China, bear live young. In oviparous species, eggs with a pliable, parchment-like shell are deposited in purpose-built burrows which are then filled in and often painstakingly concealed with leaf litter and other debris. Females may dig multiple test burrows before finding a suitable site.

In Australia, relatively open sites are often selected for egg-laying burrows, including road-verges or a pile of newly delivered topsoil for bearded dragons (*Pogona* spp.) and the bare slope of a desert dune for the Cane Grass Dragon (*Diporiphora winneckei*) (SW pers. obs). Southern Angle-headed Dragons (*Lophosaurus spinipes*), which live in a shaded environment, will seek out more open areas, often laying eggs in burrows beside vehicular access tracks and walking trails. It seems likely that a break in the canopy from a tree fall, would also be appealing.

Like all lizards, hatchling dragons are fully capable of fending for themselves from the moment they emerge. Clutch sizes vary, but eggs are generally more numerous in large species. Some of the bearded dragons (*Pogona* spp.) produce the biggest clutches, commonly more than 20 eggs, with reports of up to 41 (Harlow 2004) and even a husbandry account of 68 for the Central Bearded Dragon (*P. vitticeps*) (Brown 2012). Among the other large dragons, the Water Dragon (*Intellagama lesueurii*) can lay up to 17 eggs, but usually lays 6–12 (Aland 2008), and the Frill-necked Lizard (*Chlamydosaurus kingii*) lays 4–13 (Shine & Lambeck 1989). The smaller, slender *Diporiphora* spp. have recorded clutch sizes of 1–8 (Brown 2012).

The reproductive season generally begins between September and October. The males leave their over-winter shelter, and when the females emerge about two weeks later it is prime time for males to display, showing their colours and stylised postures to potential rivals and mates.

Females often mate with multiple males. Female Mallee Dragons (*Ctenophorus fordi*) mate repeatedly and apparently indiscriminately, with several partners in succession (Olsson 2001b), and females of numerous other species have been observed mating with several partners. As a result, multiple paternity, where a single clutch of eggs is fathered by more than one male, is believed to be common in dragons. Genetic research has demonstrated this in the Ornate Dragon (*C. ornatus*), where about 25 per cent of clutches result from more than one male (LeBas 2001).

Reproductive timing is often extremely variable across the continent, with species in warmer areas extending the breeding season over a longer time period. Mating in cooler regions occurs through

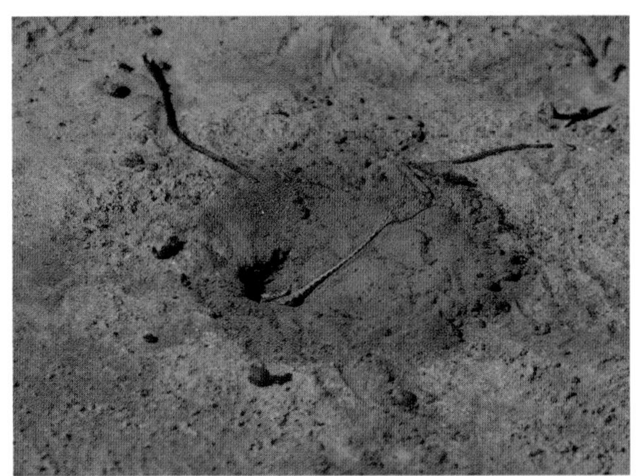

opposite page:

A pair of Thorny Devils (*Moloch horridus*) are about to mate. Female Thorny Devils are larger and stouter than males, in contrast to most dragons which have larger males. Southern NT

Photo: S. K. Wilson

this page:

Dragons bury their eggs in purpose-built nesting burrows, often sited in open areas. This Mallee Sand Dragon (*Ctenophorus fordi*) is busy excavating her burrow (top and middle) before proceeding to lay her eggs (bottom). After laying, the burrow will be filled in and rendered virtually invisible. Hattah-Kulkyne National Park, Vic

Photos: J. de Jong

spring and into early summer, with egg clutches laid from late spring into summer. Mountain Dragons (*Rankinia diemensis*) in Tasmania show a typical reproductive season for species in southerly regions, with males emerging in early September, gravid females found between October and January and the first clutch of eggs laid from October–December (Stuart-Smith *et al*, 2005). It is common for two clutches to be laid in a single breeding season. Then from mid-summer and into early autumn, hatchling dragons are a common sight.

We generally assume the sex of offspring is determined at conception by the compliment of genes received from both parents. This is the typical genetic sex determination (GSD) that applies to most reptiles (and humans too, for that matter). However, for some reptiles including crocodiles, turtles and a number of lizards including some dragons, the sex of the offspring is determined by the incubation temperature at critical periods of development. This is called temperature-dependant sex determination (TSD).

Among dragons, the means of sex determination can differ, even between closely related species. The Jacky Lizard (*Amphibolurus muricatus*) uses TSD while its closest relative, the Mallee Tree Dragon (*A. norrisi*), operates with GSD. Other dragons using TSD include the Water Dragon, Frill-necked Lizard, Ornate Dragon (*Ctenophorus ornatus*) and Painted Dragon (*C. pictus*) (Harlow, 2004).

When TSD applies, eggs incubated at high and low temperatures generally produce females while those incubated at intermediate temperatures result in varying proportions of both sexes. Critical temperature levels can vary between temperate and tropical to arid-zone species, but just a degree or two Celcius may be enough to tip the gender balance.

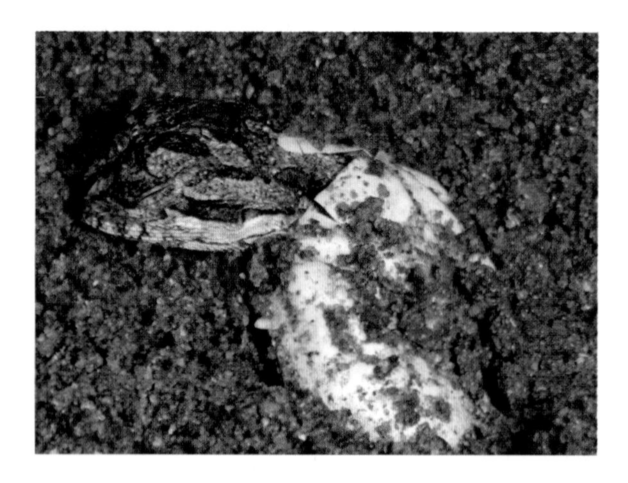

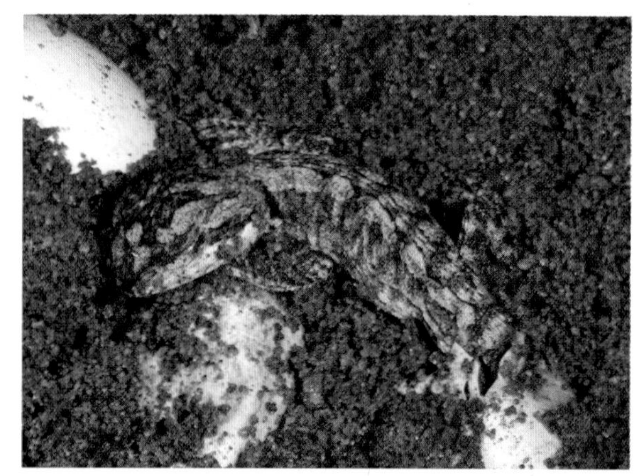

The eggs of Jacky Lizards (*Amphibolurus muricatus*) and Water Dragons, both of which occur in mild climates, produced 100 per cent females at 25°C. At 26°C, 22 per cent and 37 per cent respectively were males (Harlow & Taylor 2000; Harlow 2004). For tropical dragons, including the Frill-necked Lizards and Ta-ta Lizards (*Lophognathus* spp.), 26°C produced almost entirely females. However, this temperature is probably too low for normal embryonic development for the Frill-necked Lizard as many of the all-female offspring suffered abnormalities. For both of these species, healthy young of both sexes were incubated at 29°C (Harlow & Shine 1999).

A potential consequence of a warming climate may be a shift in the gender bias for species where sex is determined by incubation temperatures. This has been a concern of conservationists in relation to sea turtles, and appears to now be occurring among Green Turtles (*Chelonia mydas*) in the northern Great Barrier Reef. It is worth also considering its effects on lizards including dragons. There are likely more species than is currently recorded, busy digging their nest holes in exposed ground, unaware that the gender balance of their offspring may be undergoing a radical and permanent change.

opposite page:
Dragons' egg shells, like those of most lizards, are pliable and parchment-like. This Eastern Bearded Dragon (*Pogona barbata*) does not crack the shell like a bird, but slits it using an appendage called an egg-tooth on the snout. The malleable shell is then torn as the hatchling struggles free. Esk, Qld
Photos: S. K. Wilson

Dragons must keep a keen eye out for danger from all quarters. Because dragons are diurnal and often seek exposed perches, they are perhaps more susceptible to attacks from the air than other lizards. A Lozenge-marked Dragon (*Ctenophorus scutulatus*) has fallen victim to butcherbirds. Goongarrie Station, WA

photo: M. Binns

# SURVIVAL

The fight for survival is constant, with ongoing battles for food, avoiding predators, or overcoming disease and parasite infection. This struggle is interwoven into every facet of life. Unwary dragons provide tasty food items for numerous predators and spend much of their life trying to avoid being eaten.

Birds of prey frequently swoop down to pluck dragons off their perches, so it is no wonder they constantly scan the skies for signs of airborne danger. Other predators known to regularly take dragons include quolls (Glen & Dickman 2006), monitors, larger dragons (eg, Mayes *et al*, 2005, SW pers obs) and snakes (Fitzgerald *et al*, 2004). Even dragon eggs can provide a tasty meal for a hungry snake (Trembath *et al*, 2009).

Birds of prey, goannas and snakes are native predators; more recently, feral predators, particularly cats and foxes, are taking a heavy toll. Feral cats are skilled and efficient lizard hunters and are known to prey on lizards both small and large. In north-eastern Australia, lizards – including a range of dragons of the genera *Ctenophorus*, *Lophognathus*, *Pogona* and *Tympanocryptis*—comprised 41 per cent of the diet during a study of feral cats (Kutt 2011). Many millions of lizards are killed annually by cats.

Dragons employ different tactics to avoid predation. The use of pattern and colour as camouflage has already been discussed but dragons also use a range of behaviours to escape predators. Some dragons stay close to shelter—often vegetation, a rock or a safe burrow—to escape from predators. Central Netted Dragons (*Ctenophorus nuchalis*), for example, often have several burrows so they can select the nearest for a short dash to safety.

Many other dragons use high-speed evasion as a quintessential element in the survival kit. A Crested Dragon (*Ctenophorus cristatus*) sprinting away on its hind legs in a semi-arid woodland or an Ornate Dragon (*C. ornatus*) racing on all fours across a bare sheet of granite are little more than blurs to the human observer. Rufus Sand Dragons (*C. rubens*) are arguably one of the fastest Australian lizards for their size. They are impressive to behold, albeit briefly, as they streak across bare sand between spinifex clumps, make right-angle turns without any apparent decrease in velocity, then vanish from sight over a sand swale. Young Water Dragons (*Intellagama lesueurii*) can flee across water, running bipedally over a rocky stream, and they can also dive into the water from a great height. Water dragons of all ages can remain submerged for several minutes until the threat has gone.

Some dragons freeze rather than run from a threat. The pebble-mimicking earless dragons in the *Tympanocryptis cephalus* species group are the masters of this behaviour—their pebble-like shape and colouring make them almost impossible to see. A different tactic is used by the forest dragons

opposite page:
Goannas are significant predators of dragons, excavating their eggs and pursuing active lizards. This Gould's Monitor (*Varanus gouldii*) has made short work of a Central Netted Dragon (*Ctenophorus nuchalis*). Barkly Station, NT
Photo: S. K. Wilson

top:
A Black-headed Python has climbed a tree and captured a sleeping dragon (*Lophognathus* species)
Photo: J. Wright

bottom:
A dash for safety. Most dragons are extremely swift and some, like the Water Dragon (*Intellagama lesueurii*) can sprint on their hind limbs. Water Dragons can also dive into water to escape danger, and remain submerged for many minutes. Avon River, Vic
Photo: D. Paul

(*Lophosaurus* spp.). They avoid detection by quietly sliding around the back of the tree they are clinging to. When perched on slender trunks, their knees are all that remain visible.

Disease and parasites also play a part in their demise. It is not uncommon to find dragons infected with ticks or mites. Ticks, including kangaroo ticks, are known to infest a range of species, including Tawny Dragons (*Ctenophorus decresii*) (Radwan *et al*, 2014) and Eastern Bearded Dragons (*Pogona barbata*) (Doube 1975). Heavy infestation from these external parasites imposes a toll, with dragons likely to lose condition and their ability to reproduce successfully.

However, it is not only external parasites that infect Australian dragons. Biting flies, mosquitoes, ticks and mites also have the potential to spread blood parasites from one lizard to another. A range of these have been found in dragons, including microsporidian parasites – single celled parasites related to fungi – in Central Bearded Dragons (*P. vitticeps*) (Sokolova *et al*, 2016) and single-celled protozoa in Water Dragons (Mackerras 1961). Frill-necked Lizards (*Chlamydosaurus kingii*) are frequently infected by mosquito-borne filarial blood parasites, which are thread-like roundworms (Christian & Bedford 1995). Although these infections sound dire, it is not yet known to what extent these parasites cause chronic harm, if at all.

If a lizard survives all these threats, natural aging (senescence) will come into play and lizards will eventually die of natural causes—old age. Although natural life span is not wholly dependent on body size in Australian dragons, larger species are more likely to live longest. Water Dragons are known to survive more than 25 years in captivity, and the Frill-necked Lizard, one of the largest Australian dragons, has been shown to live at least 4–6 years in the wild, with females reaching sexual maturity at 18 months (Griffiths 1999).

Many Australian dragons are 'annual' species, meaning they live only for a single breeding season in the wild. Most of those are in the genus *Ctenophorus*, though there may be also some among the small *Diporiphora* species. As in humans, free radicals are harmful to lizards, aging their cells more rapidly. In the small short-lived dragons like the Painted Dragon (*C. pictus*), females that lay more clutches of eggs are potentially under high levels of physiological stress and greater DNA damage (Olsson *et al*, 2012a). Thus, the process of aging in these small dragons can be sped up, and is probably an integral part of their short life span.

Long-term survey work in the Simpson Desert has shown that the Central Netted Dragon (*Ctenophorus nuchalis*) and the Central Military Dragon (*C. isolepis*) have annual life cycles, with adults predominating during the breeding season in spring and summer and mainly juveniles occurring in other seasons (Dickman *et al*, 1999). This work has found that these are boom-bust species, with the amount of rainfall dramatically influencing survival, growth, clutch size and hatching success. In dry years with sparse vegetation, the Central Netted Dragon (*C. nuchalis*) is more abundant, while in wet years with more than 20 per cent vegetation cover, the Central Military Dragon (*C. isolepis*) benefits. An annual boom-bust life cycle would be of advantage in a desert environment with unpredictable rainfall.

Mallee Dragons (*Ctenophorus fordi*) senesce and die before their skulls have even reached a full level of maturity. These active little lizards burn out and die before they have completed their morphological development. It really is life (and death) in the fast lane.

The lifespans of dragons vary enormously. Some of the largest species, such as the Water Dragon (*Intellagama lesueurii*) (bottom right) survive for up to 20 years or more. At the other extreme there are annual species. Few Mallee Sand Dragons (*Ctenophorus fordi*) (top right) live longer than one year. Mt Coot-tha, Qld. Menzies area, WA
Photos: S. K. Wilson

# CONSERVATION

Many Australian dragons have vast distributions and are extremely common throughout their ranges. There are countless millions of Central Netted Dragons (*Ctenophorus nuchalis*) occupying a huge array of arid habitats across the continent; Central Bearded Dragons (*Pogona vitticeps*) are widely distributed, successful in highly modified landscapes and are a global sensation as pets and laboratory animals; and Jacky Lizards (*Amphibolurus muricatus*) are abundant in timbered areas of the east coast, from Victoria to southern Queensland. Some dragons even thrive in the human environment.

In the busy cities of Sydney and Brisbane, Water Dragons (*Intellagama lesueurii*) thrive in parks and gardens, along suburban creeks and even in parts of the central business districts. Inner Brisbane populations, occupying managed parks and gardens in a sea of commerce and traffic, have even begun to evolve separate characteristics. Measurable genetic and morphological differences have been identified between four isolated populations living between one and five kilometres apart in the heart of the city. They are actually evolving within the city environment (Littleford-Colquhoun *et al*, 2017).

However, other species have very restricted distributions. Some are ecological 'specialists', only occurring in one specific habitat type, with evidence of declining numbers due to various threats: human activities, such as agriculture, urban development, mining and altered fire regimes; feral animals, including predatory cats and foxes; and grazing animals such as rabbits, goats, donkeys and camels. These declining dragon species are of concern, and their long-term survival may depend on some level of conservation management.

The dragons that inhabit the native grasslands of eastern Australia are under threat of extinction. In south-eastern Australia, the temperate native grasslands are highly fragmented due to urbanisation and agriculture. Less than 5 per cent remains, and much of this is severely modified (Melville *et al*, 2014). Several species of earless dragons restricted to these grasslands are of conservation concern. The Victorian Grassland Earless Dragon (*Tympanocryptis pinguicolla*), the Condamine Earless Dragon (*T. condaminensis*) and the Roma Earless Dragon (*T. wilsoni*) are listed as Endangered on the International

left:
The Condamine Earless Dragon (*Tympanocryptis condaminensis*) is an endangered species, restricted to a tiny area of heavily utilised agricultural land in the Darling Downs, Qld. Its habitat now consists of narrow grassy road verges and crops such as sorghum and chickpea. Bongeen, Qld
Photo: S. K. Wilson

Habitat loss and degradation are among the most critical conservation issues, not just for dragons but for an appalling number of animals and plants world-wide. Moranbah area, Qld
Photo: S. K. Wilson

Union for the Conservation of Nature (IUCN) Red List of Threatened Species (Melville et al. 2017a, 2017c and 2018a). It is possible that *T. pinguicolla*, once common on the grassy basalt plains west of Melbourne, is now extinct.

The long-term prospects for species with highly restricted distributions are tenuous. A local event—such as mining, a fire or grazing pressure—could have devastating consequences. The Lake Disappointment Dragon (*Ctenophorus nguyarna*), named after a large salt lake in Western Australia, is known only around its north-eastern bank. The area falls within a proposed mining venture, and infrastructure development is likely to impact part of its habitat (Catt *et al*, 2017).

Ideally, highly specialised species with narrow habitat options and restricted distributions—those most at risk—should be noted and adequately protected. Potential threats need to be identified and measures taken before the species declines and requires intensive conservation management. The Kimberley region of Western Australia, has a number of such species, including the Gravel Dragon (*Cryptagama aurita*), and the Pale Two-pored Dragon (*Diporiphora pallida*). Not much is known about these species, making the job of conservation even more difficult. The area is also home to one of Australia's most enigmatic lizards. The Crystal Creek Two-lined Dragon (*D. convergens*) is known from just one specimen collected in 1972. Its conservation status is unknown.

The conservation management of the grassland earless dragons (*T. lineata*, *T. osbornei* and *T. condminensis*) has been tackled on a number of fronts, including population monitoring, grassland management and public education. But for other species such active management is yet to be developed. The search continues for the Victorian Grassland Earless Dragon (*T. pinguicolla*), not seen alive since the 1960s.

We can all play a role in the future of our amazing biodiversity, whether it is through making changes to minimise our impact on species or advocating their conservation. In an increasingly urbanised society, we need to reinforce the value of natural ecosystems to new generations in our schools. And politicians also need to ensure that correct evidence-based decisions are made and acted upon with appropriate funding.

With careful management, dragons will continue to display their crests, frills and dewlaps, sport their gaudy colours and live their fascinating lives across the Australian landscape for countless decades into the future. We owe it to them, and our collective wellbeing is enhanced by the knowledge that they are still out there doing their thing.

# GLOSSARY & QUICK GUIDE TO GENERA

DRAGON LIZARDS OF AUSTRALIA

Lake Eyre Dragon
*Ctenophorus maculosus*
photo M. Clancy

**Illustration 1**
**HEAD**
LATERAL VIEW

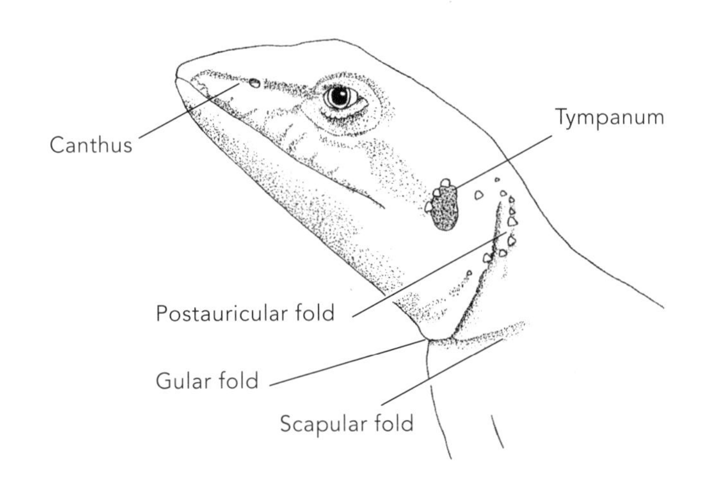


| | |
|---|---|
| **Bipedal:** | In dragons, pertains to standing or running on the hind limbs |
| **Canthal stripe:** | Linear marking along the angle of the snout anterior to the eye |
| **Canthus rostralis:** | Ridge, either sharply or gently rounded on the juncture of the side of the snout and the top of the head. **Illustration 1** |
| **Cloaca:** | Single external body opening for faeces, urine and reproduction. Present in amphibians, reptiles, birds, elasmobranch fishes (sharks and rays), and monotremes (platypus and echidnas). **Illustration 2** |
| **Dewlap.** | Extendable fold of skin on the throat |
| **Dorsal:** | Pertaining to the back or top of an animal |
| **Dorsolateral:** | The region along the junction between the back and sides, usually in reference to a longitudinal stripe or row of distinctive scales. **Illustration 3** |
| **Femoral pores:** | Small openings located in enlarged scales along the underside of the thigh, usually better developed in males. They are plugged with a wax-like substance. **Illustration 2** |
| **Granular scales:** | Very small, even, juxtaposed scales |
| **Gular:** | Pertaining to the throat |
| **Gular fold:** | Fold of skin across the throat. If obscure, it may be determined by a change in scale size. It is a diagnostic feature on dragons of the genus *Diporiphora*. **Illustration 1** |

# GLOSSARY

**Illustration 2**
**VENTRAL SURFACE**

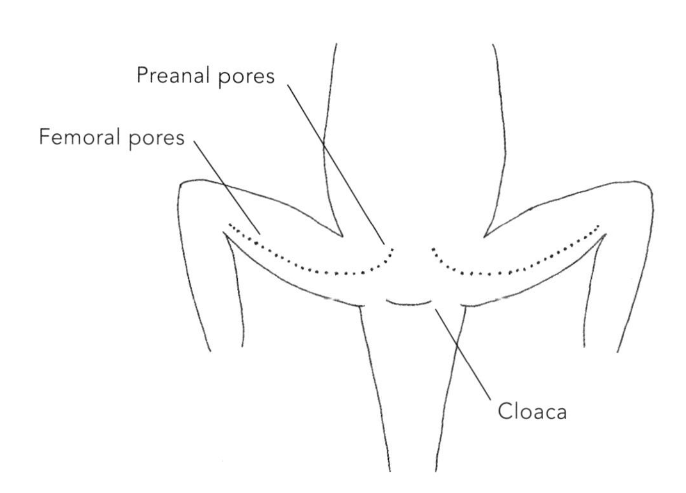


**Heterogeneous scales:** Scales that are mixed in size, shape or presence or absence of keels. Uneven scales

**Holotype:** The original specimen, lodged in a museum, on which a species' description is based and to which the scientific name is permanently fixed

**Homogeneous scales:** Scales that are uniform in size, shape or presence or absence of keels. Even scales

**Hyoid bones:** A complex of bones, associated with cartilaginous structures, which normally function to control the tongue. In iguanians they also contribute to expanding and erecting the throat and any skin folds or pouches that occur there

**Imbricate scales:** Scales that overlap

**Juxtaposed scales:** Scales arranged side by side, without overlapping

**Keel:** A raised ridge, usually in reference to a longitudinal keel on a scale

**Lateral:** Pertaining to the sides of the head or body

**Monotypic:** Refers to a genus containing just one species

**Nuchal:** Pertaining to the back of the neck, or nape

**Nuchal crest:** Raised line of enlarged, usually spinose scales on the midline of the nape **Illustration 3**

**Oviposition:** The laying of eggs

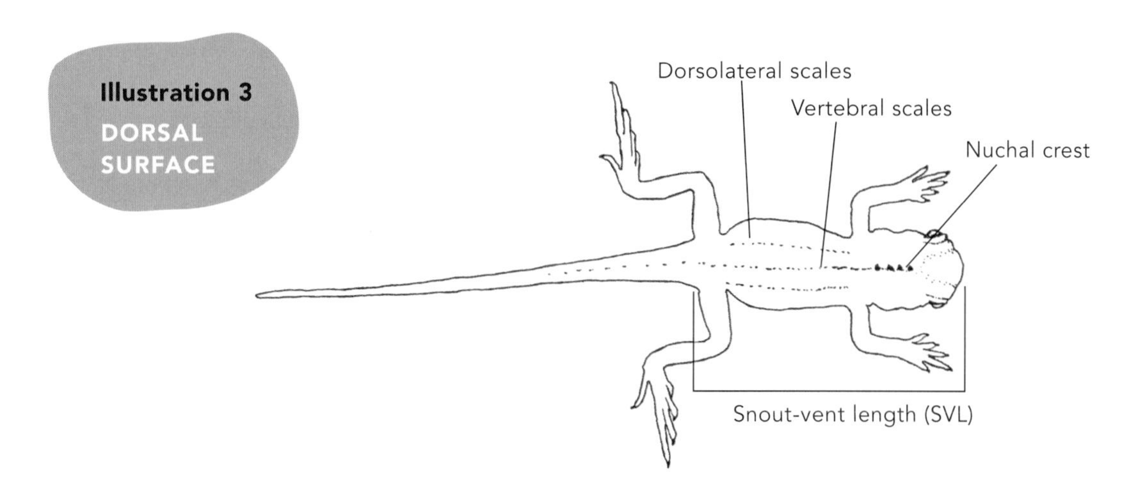


**Paravertebral:** Linear zone beside the vertebral line, usually in reference to a longitudinal stripe or row of distinctive scales

**Preanal pores:** Small openings located in enlarged scales underside of the body in front of the cloaca, usually better developed in males. They are plugged with a wax-like substance **Illustration 2**

**Postauricular fold:** Fold of skin behind the ear. It is a diagnostic feature on dragons of the genus *Diporiphora* **Illustration 1**

**Sexual dichromatism:** Difference in colouration between the sexes, often in relation to brightly coloured males or acquisition of breeding colours

**Sexual dimorphism:** Differences in morphology between the sexes, often in relation to larger crests or body size in males

**Scapular fold:** Fold of skin on the shoulder. If obscure, it may be determined by a change in scale size. It is a diagnostic feature on dragons of the genus *Diporiphora* **Illustration 1**

**Snout-vent length (SVL):** Distance between the tip of the snout and the cloaca. This is a standard measuring point in lizards. **Illustration 3**. In the following field guide we provide the average adult SVL for each species in millimetres. We also provide quick visual guide to lizard body size with a lizard silhouette for each species in comparison to a standard human hand size (19 cm length)

# GLOSSARY

**Spinose scales:** Scales with acute raised tips forming spines

**Transverse:** From side to side, as opposed to longitudinally or obliquely. Used in reference to scales, tubercles or markings

**Tubercle:** A projection from the skin. In dragons is usually refers to a scale raised above those adjacent to it

**Tympanum:** Ear drum. This is exposed in most dragons

**Ventral:** Pertaining to the underside of an animal **Illustration 1**

**Vertebral crest:** Raised line of enlarged, usually spinose or strongly keeled scales along the vertebral line of the body

opposite page
top:

*Amphibolurus burnsi.*
Bendidee Nat Park, Qld
photo R.Glor

bottom:

*Chelosania brunnea.*
Dampier Peninsula, WA
photo S. K. Wilson

# QUICK GUIDE TO GENERA

While some Australian agamid genera are distinctive and easily identified based on a suite of fixed characters, others can be problematic. The large and diverse genera *Ctenophorus* and *Diporiphora* are particularly difficult to clearly define as each contains several species that are morphologically divergent in some aspects. Where appropriate we have drawn attention to these exceptions to the rules.

The following pointers will help assign most individuals of most species to a genus. However, it should be noted that some diagnostic features such as spiny crests, appendages and body proportions and skin folds are most apparent on mature adults and may be difficult to discern on juveniles.

## AMPHIBOLURUS

### Tree Dragons

- Tympanum visible
- Nuchal crest of enlarged spines
- Three to five dorsal rows of enlarged spines; a vertebral, dorsals and dorsolaterals
- Usually some enlarged scales on upper surface of thigh, either scattered or in a row along rear edge*
- Pattern dominated by a pair of pale dorsal stripes or equivalent line of blotches

*Scales on upper surface of thigh are uniform on *A. centralis*

## CHELOSANIA

### Chameleon Dragon

- Tympanum visible and triangular
- Body laterally compressed
- Relatively short appendages, including blunt-tipped tail scarcely more than 1.5 times SVL
- Series of oblique furrows on side of neck

## CHLAMYDOSAURUS

### Frill-necked Lizard

- Tympanum visible
- Large ruff or frill folded loosely over shoulders, erectable to almost fully encircle head.
- Long slender appendages

## CRYPTAGAMA

### Gravel Dragon

- Tympanum visible
- Dorsal surface with scattered enlarged blunt tubercles
- Upper labial scales form fringe along upper lip
- Limbs and digits very short, with blunt-tipped tail shorter than SVL
- Small size: Maximum SVL = 45 mm

## CTENOPHORUS

### Comb-bearing Dragons

- Tympanum usually visible*
- A row of enlarged scales curves under eye
- Varying amounts of dark ventral pigment on mature males
- Spines and crests variable; usually absent or restricted to a line on nape, and sometimes along skin folds on sides of neck **

* Tympanum is not visible on three species: *C. maculosus* differs from *Tympanocryptis* in lacking any scattered enlarged tubercles on back, and is restricted to a salt-crust habitat; *C. parviceps* and *C. butlerorum* differ from *Tympanocryptis* in having a 'terraced' chin, where a straight line forms an angular junction along ventrolateral edge of jaw. They are restricted to west coastal sand and shell grit while the nearest *Tympanocryptis* occupy hard stony or clay based flats

** *C. cristatus* has an additional dorsolateral row of spines. *C. adelaidensis, C. chapmani, C. butlerorum* and *C. parviceps* have enlarged scattered tubercles on back

opposite page (top to bottom):

*Chlamydosaurus kingii.*
Mitchell Plateau, WA
photo S. K. Wilson

*Cryptagama aurita.*
Wave Hill, NT
photo P. Horner

*Ctenophorus butlerorum,*
Tamala Station, WA
photo B. Maryan

this page (top to bottom):

*Ctenophorus cristatus.*
Lake Cronin, WA
photo S. K. Wilson

*Ctenophorus isolepis.*
Hay River, NT
photo S. K. Wilson

*Ctenophorus maculosus.*
Lake Eyre South, SA
photo S. K. Wilson

*Ctenophorus slateri.*
Morney Station, Qld
photo S. K. Wilson

## DIPORIPHORA

### Two-lined Dragons

- Tympanum visible
- Limbs and tail usually relatively long*
- Dorsal scales extremely variable including: homogeneous with no enlarged crests (eg. *D. ameliae* & *D. lalliae*); heterogeneous with vertebral and several adjacent rows enlarged and strongly keeled (eg. *D. bilineata*); strongly heterogeneous with nuchal crest of spines and three to five greatly enlarged dorsal rows (eg *D. nobbi* and *D. amphiboluroides*)
- Keels on dorsal scales usually parallel to midline**
- Pattern, when present follows several broad trends: a pair of pale dorsal stripes sometimes overlaying dark transverse bars; or a dark shoulder patch; or a combination of both

*Appendages relatively short on *D. bennettii*
** Keels on dorsal scales converge back towards midline on *D. convergens* and *D. paraconvergens*

this page (from top to bottom):

*Diporiphora ameliae.*
Valetta Station, Qld
photo S. K. Wilson

*Diporiphora magna*
Top Springs, NT
photo R. Glor

*Diporiphora nobbi.* Bowling Green Bay National Park, Qld
photo S. K. Wilson

opposite page (from top to bottom):

*Gowidon longirostris*
Ormiston Gorge, NT
photo R.Glor

*Intellagama lesueurii.*
Brisbane River, Qld
photo S. K. Wilson

*Lophognathus horneri.*
King Edward River, WA
photo S. K. Wilson

## GOWIDON

### Long-nosed Dragon

- Tympanum visible
- Long slender limbs and extremely long slender tail
- Long snout
- An erectable crest of small spines on neck
- Only one row of enlarged scales on back, running along vertebral line
- Keels on dorsal scales converge back towards midline
- Pattern dominated by simple pale dorsal stripes, discontinuous or narrowly continuous with pale stripe along lower jaw

## INTELLAGAMA

### Water Dragon

- Tympanum visible
- Body weakly laterally compressed
- Tail strongly laterally compressed
- Prominent spiny nuchal crest continuous with vertebral crest and extending onto tail
- Large size: up to SVL 200mm

## LOPHOGNATHUS

### Ta-Ta Lizards

- Tympanum visible
- Moderately long snout
- Long slender limbs and tail
- An erectable crest of small spines on neck continuous with vertebral row of enlarged scales
- One or two additional dorsal rows of slightly enlarged scales
- Keels on dorsal scales parallel to midline
- Pattern dominated by simple pale dorsal stripes

## LOPHOSAURUS

### Rainforest Dragons

- Tympanum visible
- Body strongly laterally compressed
- Tail round in cross-section
- Conspicuous nuchal crest comprising a raised ridge and large compressed spines
- Prominent vertebral crest of spines
- Transverse gular fold and extendable dewlap present

## MOLOCH

### Thorny Devil

- Robust body with very short limbs and tail
- An armoury of thorn-like spines on head, body and appendages
- A two-spined hump on neck

top:
*Lophosaurus boydii.*
Mossman Gorge, Qld
photo S. K. Wilson

bottom:
*Moloch horridus*
Great Victoria Desert, WA
photo J. Melville

## POGONA

### Bearded Dragons

- Tympanum visible
- Body dorsoventrally compressed
- Dorsal surface with scattered enlarged raised tubercles
- One or more rows of enlarged slender spines along flanks
- A transverse row of dorsal spines across rear of head
- Usually a row of enlarged spines across throat, on an erectable spiny 'beard'

## RANKINIA

### Mountain Dragon

- Tympanum visible
- Dorsal surface with scattered enlarged raised tubercles
- Limbs and digits relatively short
- An enlarged series of spines along either side of tail-base
- Pattern dominated by two broad pale dorsal stripes with straight outer edges and deeply zig-zagging inner edges

top:
*Pogona barbata*. Brisbane, Qld
photo S. K. Wilson

bottom:
*Rankinia diemensis*, Grampians, Vic
photo D. Paul

## TROPICAGAMA

### Swamplands Lashtail

- Tympanum visible
- Long slender limbs and extremely long slender tail
- Long snout
- An erectable crest of small spines on neck.
- Only one row of enlarged scales on back, running along vertebral line
- Keels on dorsal scales converge back towards midline
- Pattern dominated by simple pale dorsal stripes, broadly continuous with pale stripe along lower jaw

## TYMPANOCRYPTIS

### Earless Dragons

- No visible tympanum; completely covered by scales
- Dorsal surface with scattered enlarged raised tubercles
- Limbs and digits usually relatively short
- Small size: Maximum SVL = 74 mm

top:
*Tropicagama temporalis.*
Casuarina, Darwin, NT
photo Ross Coupland

bottom:
*Tympanocryptis tetraporophora.*
Nilpena Station, SA
photo S. K. Wilson

*Ctenophorus cristatus.*
Lake Cronin, WA
photo S. K. Wilson

# FIELD GUIDE

DRAGON LIZARDS OF AUSTRALIA

*Amphibolurus burnsi.*
Roma area, Qld
photo S. K. Wilson

# AMPHIBOLURUS

## TREE DRAGONS

*Genus Amphibolurus* Wagler, 1830

**DESCRIPTION:** Four described species of medium-sized dragons (up to SVL 140 mm) with moderately robust bodies, long limbs and very long slender tails. Body scales heterogeneous, including a nuchal and vertebral row of enlarged spinose scales, 1–2 similar dorsal rows, and usually additional irregular enlarged scales on back and dorsal surface of thighs. Tympanum is exposed. Typically shades of grey to black with a pair of broad pale dorsal stripes which may be continuous or constricted into a series of lozenge-shapes, and often a broad pale stripe through lips. Colour and pattern intensity can vary markedly with temperature, disposition, sex and breeding status. Femoral pores range from 2–8 and preanal pores from 3–11.

**KEY CHARACTERS:** Species differ from *Lophognathus* and *Gowidon* in having more heterogeneous scalation and further from *Gowidon* in having more dorsal rows of enlarged spinose scales (versus only the vertebral row enlarged).

**DISTRIBUTION AND ECOLOGY:** Open forests, woodlands and timbered margins of water courses of southern, eastern and central Australia. Terrestrial and semi-arboreal, often perching on stumps, rocks and fallen timber, and on trunks and lower branches of trees.

**BIOLOGY:** Tree dragons feed mainly on arthropods, but also frequently include larger items such as skinks in their diets. They are sit and wait predators that actively regulate their body temperature, with a preferred body temperature between 34–38°C, depending on the season and species. Males acquire more intense colours, with the background becoming darker to black, in contrast to the pale markings which may be a sharply contrasting white. Males also attain a greater body size and larger heads than females, which probably provides greater advantage in competition for territories and mates. Social signalling includes head bobs and dips, and tail lashing or flicking. Clutches of 6–11 eggs for most species, although smaller clutches have been recorded in *A. norrisi*. The nest burrow is often located in an exposed site. In *A. muricatus* the sex of offspring is determined by incubation temperature, while in *A. norrisi* it is genetically determined.

## BURNS' DRAGON

*Amphibolurus burnsi* Wells and Wellington, 1985

**DESCRIPTION:** SVL 110 mm. Large robust lizard. Wide head with extensive covering of spinose scales. Dorsal scales strongly heterogeneous, with well-developed spinose nuchal and vertebral crest back to hips, and two dorsolateral rows of spinose scales extending from shoulders to hips. Scales on thighs strongly heterogeneous with scattered spinose scales and a prominent row of enlarged spinose scales along the posterior edge. Shades of brown, grey to almost black. Two broad pale dorsolateral stripes running from ear or neck to hips, discontinuous with pale stripe along lower jaw. Mouth-lining pink. Femoral pores 3–5; preanal pores 4–6.

**KEY CHARACTERS:** Differs from the Centralian Tree Dragon (*A. centralis*) in having strongly heterogeneous scales on the thigh. Differs further, and from the Jacky Dragon (*A. muricatus*) in having a prominent row of spinose scales running along the posterior edge of thigh.

**DISTRIBUTION AND ECOLOGY:** Occurs in dry woodlands of the eastern interior, including communities of Poplar Box, *Callitris* pines, Brigalow and ironbark. Also associated with eucalypts along inland watercourses. Distributed across southern and central-western Qld and northern inland NSW.

**BIOLOGY:** Little known about biology of this species. Commonly seen on the ground or perching on trees and woody debris.

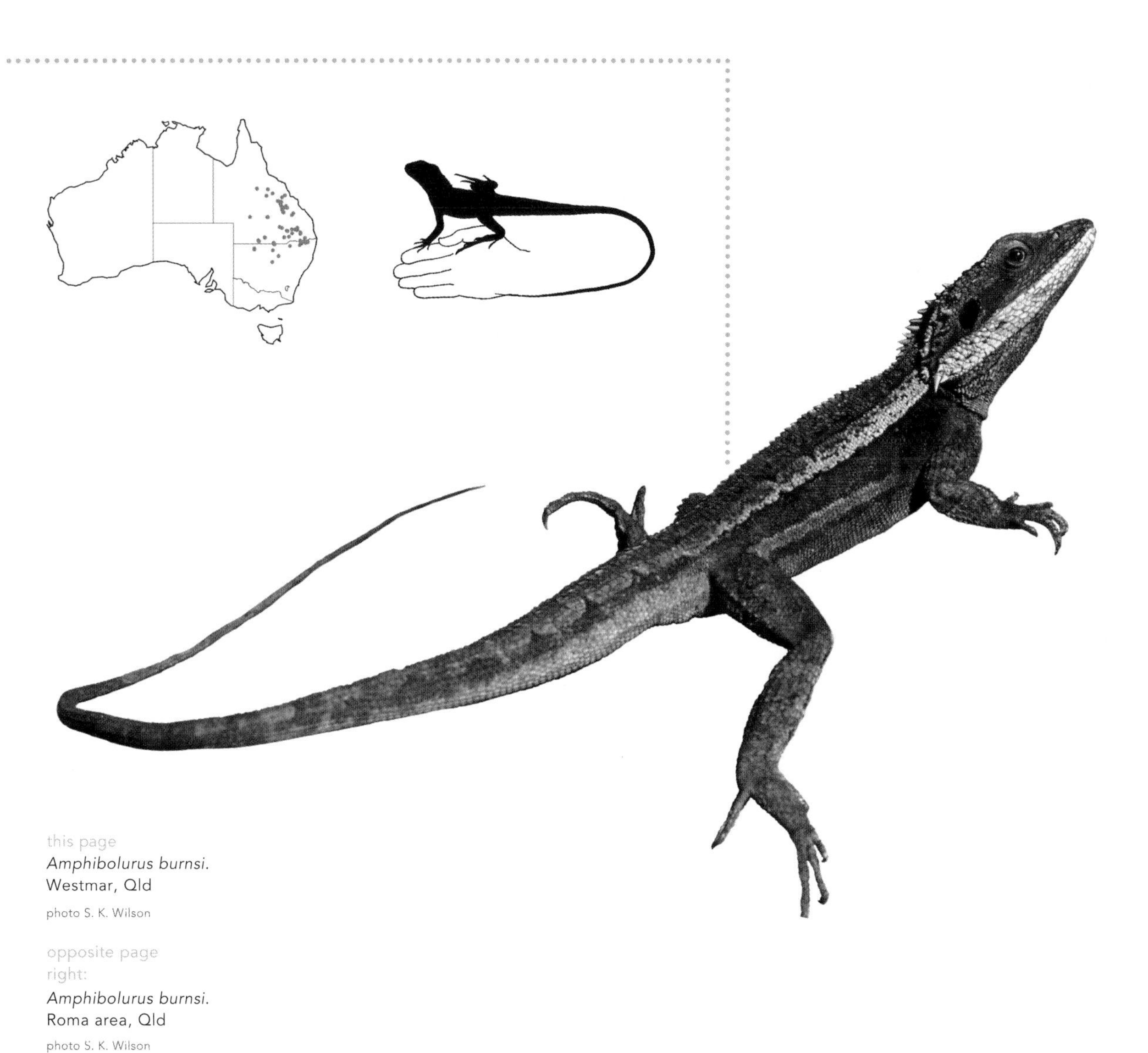

this page
*Amphibolurus burnsi.*
Westmar, Qld

photo S. K. Wilson

opposite page
right:
*Amphibolurus burnsi.*
Roma area, Qld

photo S. K. Wilson

left:
*Amphibolurus burnsi* habitat.
Yuleba State Forest, Qld

photo S. K. Wilson

## CENTRALIAN TREE DRAGON

*Amphibolurus centralis* Loveridge 1933

**DESCRIPTION:** SVL 140 mm. Robust with large, wide head in proportion to body. Well-developed nuchal and vertebral crest and 1–2 additional rows of enlarged scales on back. Scales on back heterogeneous, with keels on inner-most dorsal scales parallel to midline, those on outer enlarged rows parallel to or converging back towards midline, and keels on remaining small dorsal scales converging to midline. Scales on thighs relatively homogeneous. Light to dark brown and grey with two broad pale dorsolateral stripes running from ear or neck to the hip, discontinuous with pale lip scales. Dorsolateral stripes sometimes intersected by multiple wedges of brown or grey along their length. Most individuals have a broad pale or white stripe running along extent of the lower lip, though some individuals lack this feature. Diffuse pale stripe present between eye and ear in some individuals. Femoral pores 2–6; preanal pores 3–6.

**KEY CHARACTERS:** Differs from *A. burnsi* by lacking spinous scales on the thigh and no enlarged spinous scales along rear of thigh.

**DISTRIBUTION AND ECOLOGY:** Arid northern-central and central Australia, particularly associated with mulga woodlands but also occurring in eucalypt woodlands, open mallee and timbered margin of water courses. Western Qld, NT, and WA. Commonly seen clinging vertically to smaller trunks or in the branch forks of mulga trees.

**BIOLOGY:** Little known about biology of this species, but probably similar to *A. burnsi*. An average active field body temperature of 33.7°C has been recorded (listed as *L. gilberti*) and it was noted that they spend approximately 68% of their time in the shade when active (Melville & Schulte, 2001).

**NOTES:** Genetics research has confirmed that this species is unrelated the *Gowidon* and *Lophognathus* species, and demonstrates it is closely related to *A. burnsi* (Melville *et al*, 2011).

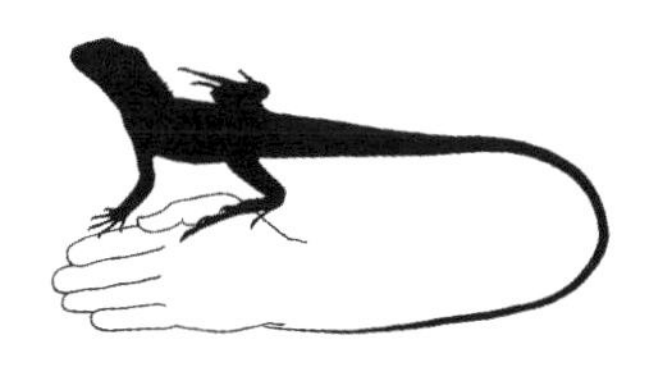

opposite page left:
*Amphibolurus centralis* habitat.
Lark Quarry area, Qld
photo S. K. Wilson

opposite page right:
*Amphibolurus centralis.*
West MacDonnell Ranges, NT
photo J. Melville

below:
*Amphibolurus centralis.*
Tennant Creek area, NT
photo S. K. Wilson

opposite page:
*Amphibolurus muricatus.*
Girraween National Park, Qld
photo S. K. Wilson

top:
*Amphibolurus centralis,*
Mature male.
Noonbah Station, Qld
photo S. K. Wilson

bottom:
*Amphibolurus centralis.*
Winton, Qld
photo S. K. Wilson

# AMPHIBOLURUS

## JACKY LIZARD

*Amphibolurus muricatus* White, 1790

DESCRIPTION: SVL 120 mm. Large and robust with well-developed nuchal and vertebral crests. Scales on back and sides strongly heterogeneous, with keels on dorsal scales parallel to or weakly converging on midline. Two rows of enlarged scales on either side of the vertebral scale ridge and usually a row of enlarged lateral scales. Thighs covered with scattered enlarged spinose scales, interspersed with small scales. No spinose scales on sides of tail-base. Slate grey to brown with dark grey to black patterning. A dark vertebral stripe and pale dorsal stripes, sometimes broken by dark transverse bars to form a series of lozenge-shaped blotches. No dark canthal stripe from tip of nose to eye. Dark transverse markings usually present across snout between nostrils. Dark stripe between eye and tympanum bordered above by a light stripe; upper and lower lips pale, extending to back of head, sometimes tinged with orange. Inside of mouth pale to bright yellow. Femoral pores 4–8; preanal pores 4–11.

KEY CHARACTERS: Differs from Mallee Tree Dragon (*A. norrisi*) in patterning, lacking a dark canthal stripe, usually having dark transverse marking on snout and usually brighter dorsal pattern. Differs from Burns' Dragon (*A. burnsi*) in lacking an enlarged row of spinose scales along rear edge of thigh. Differs from Nobby Dragon (*Diporiphora nobbi*) by having more heterogeneous dorsal scales, particularly on upper surface of thigh, and a yellow mouth-lining (versus pink). Differs from Mountain Dragon (*Rankinia diemensis*) in having a yellow mouth-lining (versus pink), lacking spinose scales on sides of tail-base and having relatively long limbs and tail (versus short limbs and tail).

DISTRIBUTION AND ECOLOGY: Widely distributed along the east coast and inland south-eastern Australia. Occurs in many habitats, including dry sclerophyll forests, open woodlands and heaths, from coastal lowlands to highlands, but absent from the alpine zone. Commonly seen in forested areas peripheral to Sydney and Melbourne.

BIOLOGY: Territorial social structure with polygynous mating. Known to use tail flicks, head bobs and arm waving in social behaviour. Clutch size of 3–9 eggs recorded, with a female often laying two clutches in a reproductive season (Oct-Feb). Sex of offspring is determined by incubation temperature (Harlow & Taylor, 2000). Juveniles hatching early in the season have been found to disperse further than those hatching later (Warner & Shine, 2008).

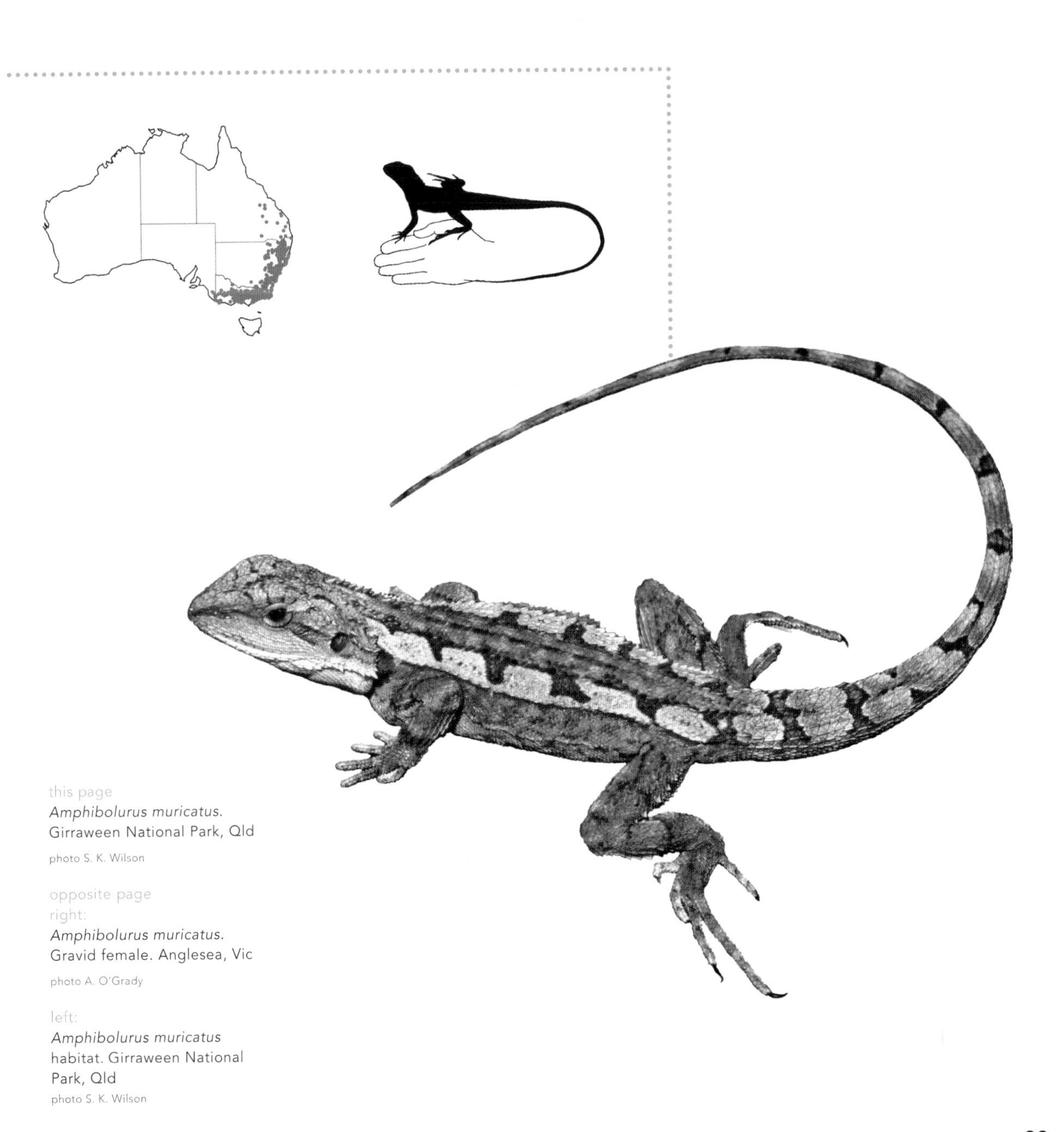

this page
*Amphibolurus muricatus*.
Girraween National Park, Qld

photo S. K. Wilson

opposite page
right:
*Amphibolurus muricatus*.
Gravid female. Anglesea, Vic

photo A. O'Grady

left:
*Amphibolurus muricatus*
habitat. Girraween National Park, Qld

photo S. K. Wilson

## MALLEE TREE DRAGON

*Amphibolurus norrisi* Witten & Coventry, 1984

DESCRIPTION: SVL 69 mm. Medium-sized, somewhat elongate with moderately-developed nuchal and vertebral crest. Scales on back and sides strongly heterogeneous, with keels on dorsal scales parallel to or weakly converging on midline. Rows of enlarged scales on back, consisting of two rows either side of the vertebral scale ridge and usually a row of enlarged lateral scales. Slate grey to brown with weak dark grey to black patterning. A dark vertebral stripe and pale dorsal stripes, with or without dark transverse bars breaking pale stripes into a series of lozenge-shaped blotches. Diffuse dark bars on base of tail, forming more distinct banding on second half. Dark canthal stripe from tip of nose to eye. Dark stripe between eye and tympanum bordered above by a light stripe; upper and lower lips pale, extending to back of head, sometimes tinged with orange. Inside of mouth pale yellow. Femoral pores 4–8; preanal pores 4–11.

KEY CHARACTERS: Differs from the Jacky Lizard (*A. muricatus*) in having a dark canthal stripe from snout to eye and usually weaker dorsal pattern. Differs from Nobby Dragon (*Diporiphora nobbi*) by having more heterogenous dorsal scales, particularly on thighs, and a yellow mouth-lining (versus pink).

DISTRIBUTION AND ECOLOGY: Semi-arid mallee woodlands and heaths from north-western Vic. to Ravensthorpe area, WA. The distribution is fragmented, and includes an isolated population from the coastal heaths and woodlands of the Eyre Peninsula to the Great Australian Bight. Populations on the Eyre Peninsula have a larger body size than those in the eastern part of the range.

BIOLOGY: Territorial social structure with polygynous mating and female-biased dispersal. Clutch size of 3–7 eggs. The sex of off-spring is determined genetically (Warner *et al*, 2009). Age estimates, based on counting growth rings in long-bones, suggests that *A. norrisi* lives for 5–7 years in the wild (Smith *et al*, 2013).

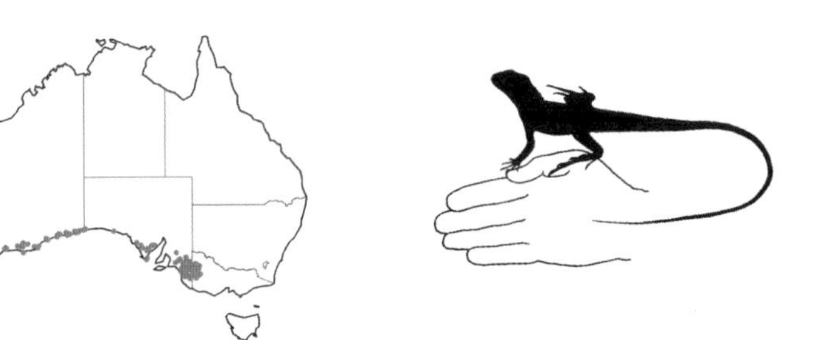

opposite page left:
*Amphibolurus norrisi* habitat.
Wagon Flat Bore area, Big Desert, Vic
photo S. K. Wilson

opposite page right:
*Amphibolurus norrisi*. Border Village, SA
photo S. K. Wilson

below:
*Amphibolurus norrisi*.
Wagon Flat Bore area, Big Desert, Vic
photo S. K. Wilson

*Chelosania brunnea*
Mitchell Plateau, WA
photo S. K. Wilson

# CHELOSANIA

## CHAMELEON DRAGON

*Genus Chelosania* Gray 1845,
*Chelosania brunnea* Gray 1845

## CHAMELEON DRAGON

*Genus Chelosania* Gray 1845, *Chelosania brunnea* Gray 1845

DESCRIPTION: Sole member of genus. SVL 118 mm. Short-limbed with a blunt-tipped tail, a strongly laterally compressed head and body, a distinctive dewlap and granular eyelids with small apertures, much like its namesake, the chameleon. The ear aperture is large and wedge-shaped. There is a low erectable nuchal crest, continuous with a vertebral ridge onto tail. Colour ranges from reddish brown to grey with dark variegations and streaks, including narrow lines radiating from the eye and broad dark bands on the tail. Femoral and preanal pores absent.

KEY CHARACTERS: The laterally compressed body, blunt-tipped tail and granular eyelids with small apertures distinguish this species from all sympatric dragons.

DISTRIBUTION AND ECOLOGY: Males are recorded on the ground during May, presumably to '... take part in territorial fighting to establish mating rights' (Trainor, 2005). Females have been seen laying eggs in the mid-dry season between July and August with a clutch size of 8 reported. One nesting female took nearly all day to compete the process of burrow excavation and concealment, selecting a site that received morning sun, but shade during the hottest part of the day (A. Dudley pers. comm.). They are arthropod feeders, with Green Ants (*Oecophylla smaragdina*) recorded as prey item.

COMMENTS: Chameleon Dragons are not listed as of conservation concern. They are rarely seen but may simply be 'difficult to find'. However, it is likely that late dry season fires and feral cats represent significant threats. Genetic and chromosomal work provides strong evidence that this species along with the Thorny Devil (*Moloch horridus*) form two separate, ancient lineages in the arid zone and are independent from the large, more recent radiation of dragons (Schulte *et al*, 2003; Hutchinson & Hutchinson, 2011).

*Chelosania brunnea.*
Dampier Peninsula, WA
photo S. K. Wilson

opposite page:
*Chelosania brunnea* habitat.
Dampier Peninsula, WA
photo S. K. Wilson

*Chlamydosaurus kingii.*
Cape York Peninsula, Qld
photo J. De Jong

# CHLAMYDOSAURUS

## FRILL–NECKED LIZARD; FRILLED LIZARD

*Genus Chlamydosaurus* Gray, 1825

*Chlamydosaurus kingii* Gray, 1825

# CHLAMYDOSAURUS

## FRILL-NECKED LIZARD; FRILLED LIZARD

*Genus Chlamydosaurus* Gray, 1825, *Chlamydosaurus kingii* Gray, 1825

**DESCRIPTION:** Sole member of genus. SVL 258 mm. Large and slender with long slender limbs and long tail. This extremely distinctive lizard is characterised by a thin scaly cape lying folded over the shoulders. When erected it almost completely encircles the head. Body scales are homogeneous and there are no nuchal and vertebral crests, though scales along vertebral line are slightly larger than adjacent scales. Populations in NT and WA typically have splashes of red on their frills while those from Qld are generally shades of grey

**KEY CHARACTERS:** With its distinctive cape, the Frill-necked Lizard cannot normally be confused with any other lizard. However, it is sometimes confused with bearded dragons (*Pogona* spp.) which have an erectable spiny pouch under the throat

**DISTRIBUTION AND ECOLOGY:** Tropical woodlands across northern Australia, from just south of Broome in WA to Brisbane, Qld in the east, favouring well-drained soils that support diverse tree species. Frill-necked Lizards are most obvious during the wet season, when they perch on vertical trunks of rough-barked trees, positioning themselves at heights of 1–2 metres, and often descend to the ground. At these times they are a common sight along Top End road edges. During the dry season they tend to remain higher in the canopy.

**BIOLOGY:** This is undoubtedly one of the most recognizable lizards in Australia, having graced the two cent coin and virtually every picture book on Australian wildlife. The remarkable frill is erected as a warning to potential predators and rivals. At rest it is inconspicuous but when threatened or provoked the frill is dramatically erected to stand out at right angles as a ribbed disc nearly 30 cm across. It is supported by several long, slender hyoid bones that can be raised and lowered as the mouth is opened. The Frill-necked Lizard is also unusual in being fully bipedal. It not only runs on the hind limbs like many dragons, but it also walks in this manner. Clutches of 4–13 eggs are laid during the wet season. The sex of offspring is determined by incubation temperatures (Harlow & Shine, 1999) and genetic research suggests that juvenile males disperse further than females (Ujvari *et al*, 2008). Frill-necked Lizards feed almost exclusively on arthropods, with insects contributing over 95% of all records (Shine & Lambeck, 1989). Caterpillars constitute the largest biomass of prey ingested though Green Ants (*Oecophylla smaragdina*) and termites are extremely numerous as individual prey items. They eat little or no vegetation, in marked contrast to other large-bodied dragons and iguanas which, with increasing body size, trend strongly towards omnivory and herbivory.

this page
*Chlamydosaurus kingii.*
Kimberley region, WA
photo A. O'Grady

opposite page
right:
*Chlamydosaurus kingii* displaying.
El Sharana, NT
photo S. K. Wilson

left:
*Chlamydosaurus kingii* habitat.
King Edward River, WA
photo S. K. Wilson

**COMMENTS:** The most south-easterly population of Frill-necked Lizards near Brisbane has dramatically declined, a likely result of feral predators and habitat loss. Although Frill-necked Lizards are still abundant across the north, increasing frequency of intense fires at the end of the dry season has been found to impact the Frill-necked Lizard, with these fires killing most, and in some cases, all individuals in the affected areas (Brook & Griffiths, 2004) and isolating remaining populations (Ujvari *et al*, 2008).

*Cryptagama aurita*.
Wave Hill, NT
photo P Horner

# CRYPTAGAMA

## GRAVEL DRAGON

*Cryptagama aurita* Storr, 1981

# CRYPTAGAMA

## GRAVEL DRAGON

*Cryptagama aurita* Storr, 1981

**DESCRIPTION:** Sole member of genus. SVL 46 mm. A small robust dragon with short limbs, a tail shorter than SVL, and a rounded head with no canthus rostralis. Dorsal scales are heterogeneous, with very small scales mixed with enlarged rounded tubercles scattered across dorsal surface including limbs and tail. Upper labial scales form a unique denticulate fringe along upper lip. The tympanum is large and obvious. Pale reddish brown to brick red, sometimes with pale brownish grey suffusion on head and back. Femoral and preanal pores 10–18.

**KEY CHARACTERS:** Superficially similar to the genus *Tympanocryptis* but differs in having an obvious tympanum (versus completely covered by scales), a tail shorter than its SVL, a denticulate fringe along upper lip and a higher number of femoral and preanal pores.

**DISTRIBUTION AND ECOLOGY:** Known from a handful of localities across the southern and central Kimberley region, WA and western NT. These are: the Halls Creek area, Wolf Creek Meteorite Crater, Wave Hill and the north-eastern Kimberley. They are believed to occur on stony ground, with a spinifex dominated ground cover. The morphology of the animal indicates it is a pebble mimic, with the head and body resembling small stones. Based on the behaviour of other pebble mimics such as some earless dragons (*Tympanocryptis cephalus* group), the Gravel Dragon would probably crouch if disturbed, with the limbs pulled close to its sides. Like other small arid-adapted dragons it is an arthropod feeder, probably relying strongly on ants.

**COMMENTS:** On-going genetic and morphological work is resolving the mystery surrounding the evolutionary relationship of this species to other Australian dragon lizards and has clearly demonstrated that it is related to the diverse genus *Ctenophorus*.

this page
*Cryptagama aurita.*
Kimberley region, WA
photo S. Mahony

opposite page
right:
*Cryptagama aurita.*
Wave Hill, NT
photo P. Horner

left:
*Cryptoagama aurita* habitat.
Vicinity of the type locality, Halls Creek area, WA
photo S. Mahony

*Ctenophorus nuchalis.*
MacDonald's Yard, NT
photo R. Glor

# CTENOPHORUS

## COMB-BEARING DRAGONS

*Genus Ctenophorus* Fitzinger, 1843

**DESCRIPTION:** This is Australia's largest agamid genus, with 32 currently recognised species. Diagnostic characters vary: a row of enlarged scales curves from below eye to above ear; dorsal scales are generally small and uniform but scattered enlarged scales present on some species; enlarged nuchal and vertebral crests range from well-developed and sometimes erectable to absent; bodies weakly laterally compressed to strongly dorsally depressed; tympanum usually exposed (covered by scales on three species); limbs and tails long and slender to moderately short; sexual dichromatism ranges from striking (bright, multi-hued males and drab females) to undiscernible; black ventral pigment present to varying degrees on males. Femoral and/or preanal pores present on males.

**KEY CHARACTERS:** The species tend to cluster into the following morphological and behavioural groups. Some of these groups have been confirmed as evolutionary lineages based on genetics.

The *C. caudicinctus* - *C. ornatus* group are rock-inhabiting species that form an evolutionary lineage. Represented by *C. caudicinctus*, *C. infans*, *C. slateri*, *C. graafi*, *C. ornatus* and *C. yinnietharra*. These large robust members of the genus are extremely swift dragons that perch in elevated rocks and dash on all four limbs across granite sheets, boulder fields or other rocky to stony terrain. Males can have striking colouration and in a number of the species highly contrasting rings on their tails. Restricted to rock habitats from WA, through NT to western Qld – there are a couple of records from far north-western SA. They are probably the most abundant and visible dragons on outcrops in the Pilbara and south-western regions of WA.

The *C. decresii* group is an evolutionary lineage containing *C. decresii*, *C. fionni*, *C. mirityana*, *C. rufescens*, *C. tjantjalka* and *C. vadnappa*. These are also rock-inhabiting. They have moderately to strongly depressed heads and bodies, low erectable nuchal crests and a weak to absent vertebral ridge. Males of most species exhibit brilliant hues, combining reds, yellows and blues while the cryptic females and juveniles range from grey to reddish brown. Male breeding and territorial display can be elaborate, featuring the raising and lowering of the body while coiling the tail, terminating with head bobs and dips. They utilise elevated perches on outcrops, and shelter in rock crevices. These dragons occur in dry areas from northern and southern SA to adjacent NSW.

The *C. pictus* group, comprising *C. pictus*, *C. nguyarna* and *C. salinarum*, occupy sandplains, swales and the margins of salinas. They share deep blunt heads, robust bodies, relatively short limbs and tails, and nuchal crests and vertebral ridges ranging from weak and erectable to absent. They are terrestrial, occasionally utilising low perches, and they excavate U-shaped burrows at the bases of low shrubs. They are widespread in dry to arid zones of all mainland states.

The *C. reticulatus* group, comprising *C. clayi*, *C. nuchalis*, *C. reticulatus*, *C. gibba* and the enigmatic *C. maculosus* share very deep rounded heads with blunt snouts, robust depressed bodies and relatively short limbs and tails. Tympanums are

top:
*Ctenophorus slateri*.
Kalkaringi, NT
photo R. Glor

bottom:
*Ctenophorus pictus*.
Breeding male.
Ballera area, Qld
photo S. K. Wilson

exposed on all but *C. maculosus*. They are terrestrial dragons that select low perches such as rocks, stumps, earth piles and fence posts in open arid habitats. One species, *C. nuchalis*, is a conspicuous sight on road-side perches across the desert regions. They excavate burrows, normally at the bases of surface vegetation. The exception is *C. maculosus* that burrows under salt crusts on the open salinas of Lakes Eyre and Torrens.

The *C. maculatus* group, often known as sand and military dragons, consists of *C. isolepis*, *C. fordi*, *C. femoralis*, *C. maculatus* and *C. rubens*. These are slender and slightly depressed dragons with very long limbs and tails and little or no indication of spines or crests. These wholly terrestrial dragons rarely if ever select elevated perches. They inhabit the bare sandy ground between clumps of low vegetation, primarily porcupine grass and shrubs, in arid areas of all mainland states. All are extremely swift, generally sprinting a short distance before stopping to assess any perceived threat. They are constantly on the move, rarely pausing long, and are generally fast growing and short-lived. Two species studied, *C. fordi* and *C. isolepis* are mainly annual, with few adults surviving a second breeding season.

The *C. cristatus* group, consisting of *C. cristatus*, *C. mckenziei* and *C. scutulatus* have long limbs and tails, relatively long deep heads and well-developed spiny nuchal and vertebral crests. These are terrestrial inhabitants of semi-arid southern woodlands, primarily mallee or *Acacia*, and sometimes chenopod shrublands. They often perch on fallen timber, forage widely on open ground, and are very swift, running on their hind limbs with body erect. They are sometimes referred to as bicycle lizards.

The *C. adelaidensis* group, comprising *C. adelaidensis*, *C. chapmani*, *C. parviceps* and *C. butlerorum* are very small (maximum SVL 52 mm) with relatively deep heads, short limbs and tails. They have heterogeneous body scales including granular scales and numerous erect tubercles on back and flanks, and a row of enlarged spinose scales along each side of tail-base. *C. parviceps* and *C. butlerorum* have the tympanum covered by scales. They inhabit heaths and shrublands on sandy soils, mainly along the southern and western coasts and hinterland. They tend not to be very swift relative to other *Ctenophorus*, rarely select elevated perching sites, and scuttle on all four limbs to the shelter of low vegetation.

**DISTRIBUTION AND ECOLOGY:** It is not surprising, given the degree of variation, that *Ctenophorus* exhibit a wide range of lifestyles across a variety of habitats. The greatest diversity of this genus is seen in the arid and semi-arid interior of Australia, occurring across all available habitats. Some species also extend into the monsoon tropics in northern Australia, (including *C. slateri* and *C. nuchalis*) and southern temperate regions (*C. decresii*). All species are terrestrial though some frequently perch on low objects such as rocks, small timber, and tussocks.

**BIOLOGY:** Comb-bearing Dragons are mainly generalist arthropod feeders, but the sand and military dragons are known to have a high proportion of ants in their diet, while some of the larger species, such as *C. nuchalis* eat a high proportion of plant material. They are sit and wait predators that actively regulate their body temperature. Males are larger than females and acquire more intense colours, which probably provides greater advantage in competition for territories and mates, but may increase mortality from predation. Social signalling and male colouration has been well studied in some of the species groups, particularly the *C. decresii* group. Clutch sizes vary depending on the size of the species, with the smaller species having clutches of 2–4 eggs, while larger species are known to have clutches of up to 9 eggs. Eggs are laid in burrows, with most species able to produce two clutches per year. Many of the species are annual and only a small proportion of individuals survive to their second year.

# CTENOPHORUS

## WESTERN HEATH DRAGON

*Ctenophorus adelaidensis* Grey, 1841

**DESCRIPTION:** SVL 52 mm. Very small with short limbs and tail. Body scales heterogeneous; one or more erect, multi-keeled tubercles on each side of neck, dorsal body scales variable including keeled and granular scales, numerous erect tubercles on back and flanks mainly restricted to areas of dark colour, and a row of enlarged spinose scales along each side of tail- base. Chin 'terraced' (a straight line forming angular junction along ventrolateral edge of jaw). Grey with a broad vertebral stripe margined by black triangular blotches, two narrow dark dorsolateral lines and a row of angular upper lateral blotches. Ventral surfaces white with black on males or grey on females as follows: a chevron on the throat, a patch on the chest and three continuous dark ventral stripes which converge in front of the vent. Preanal pores 2–5.

**DISTRIBUTION AND ECOLOGY:** Sandplains supporting heaths and woodlands, including coastal heath of *Agonis flexuosa*, *Spyridium globulosum* and *Acacia rostellifera* shrublands on coastal sand. Distributed on the lower west coast of WA from Kalbarri south through the Perth area to south of the Swan River (Thompson *et al*, 2008). Fully terrestrial, rarely utilising elevated perches. Not particularly swift when compared to other small dragons, tending to rely mainly on cryptic disruptive pattern to avoid predators. When pursued it scuttles rather than sprints to the cover of low vegetation.

**BIOLOGY:** Ants probably form a major component of the diet.

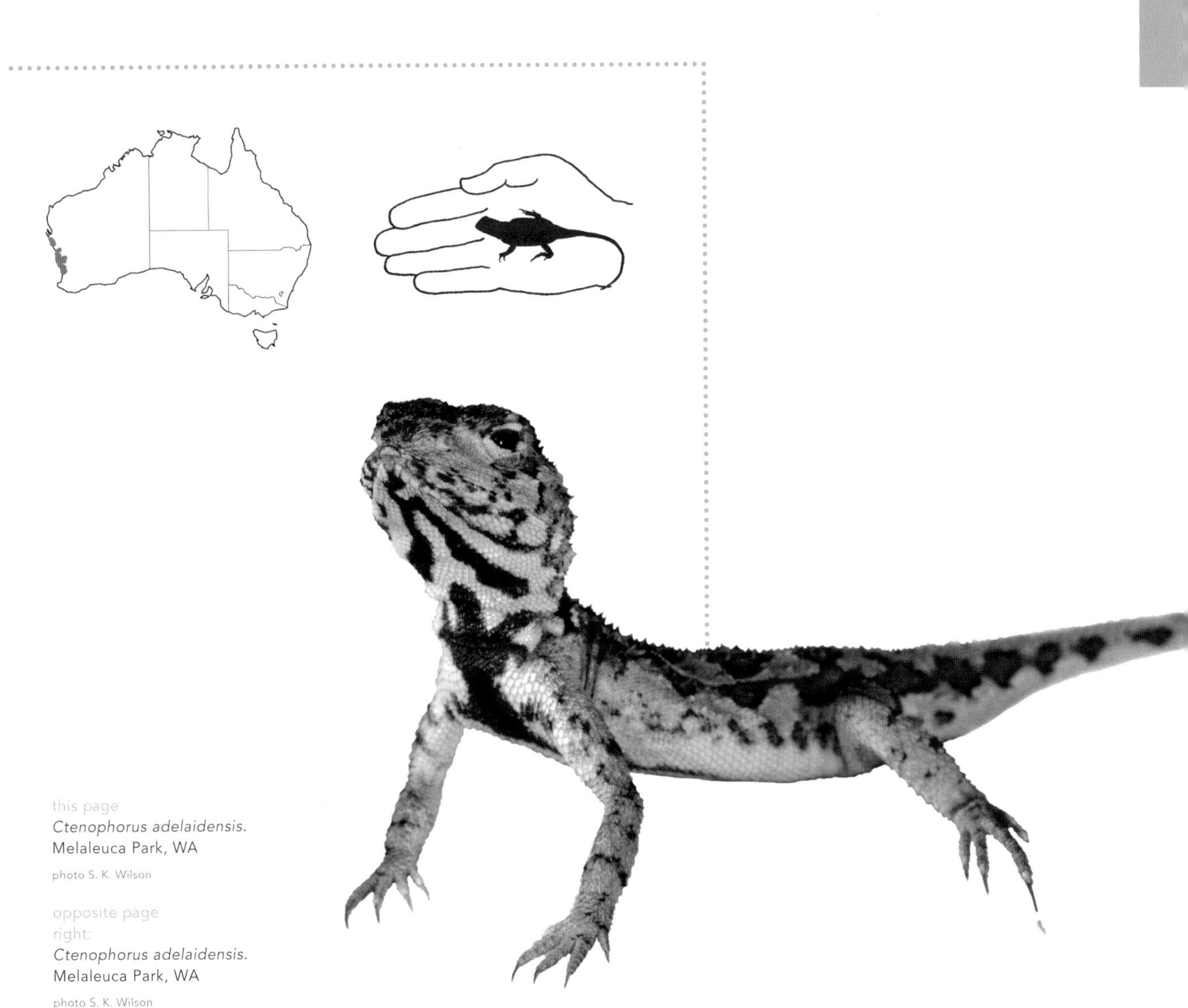

this page
*Ctenophorus adelaidensis.*
Melaleuca Park, WA

photo S. K. Wilson

opposite page
right:
*Ctenophorus adelaidensis.*
Melaleuca Park, WA

photo S. K. Wilson

left:
*Ctenophorus adelaidensis*
habitat. Coastal plain, Melaleuca Park, WA
photo S. K. Wilson

## SHARK BAY HEATH DRAGON

*Ctenophorus butlerorum* Storr, 1977

**DESCRIPTION:** SVL 43 mm. Very small with short limbs and tail. Body scales heterogeneous; uniform and weakly spinose down middle of back, but varying greatly in size on dorsolateral area, including erect tubercles mainly restricted to areas of dark colour. Tympanum completely covered by scales. Chin 'terraced' (a straight line forming an angular junction along ventrolateral edge of jaw). Pale to dark grey with a broad unbroken vertebral stripe, 8–10 dark hourglass-shaped transverse bars, dark upper flanks with darker blotches, indistinct pale midlateral stripe from base of tail nearly to forelimb and bright yellow on chin and lips. Femoral and preanal pores 14–20.

**KEY CHARACTERS:** Differs from the North-western Heath Dragon (*C. parviceps*) by its stronger pattern, yellow chin and lips, and fewer femoral and preanal pores (14–20 versus 26–34).

**DISTRIBUTION AND ECOLOGY:** Restricted to a narrow strip along the mid-west cost of WA, between Shark Bay and Kalbarri. Occupies white coastal dunes and pinkish sandplains, vegetated with heaths and beach spinifex (*Spinifex longifolius*). Not particularly swift when compared to other small dragons, tending to rely mainly on its cryptic disruptive pattern to avoid predators. When pursued it scuttles rather than sprints to the cover of low vegetation.

**BIOLOGY:** Ants probably form a major component of the diet.

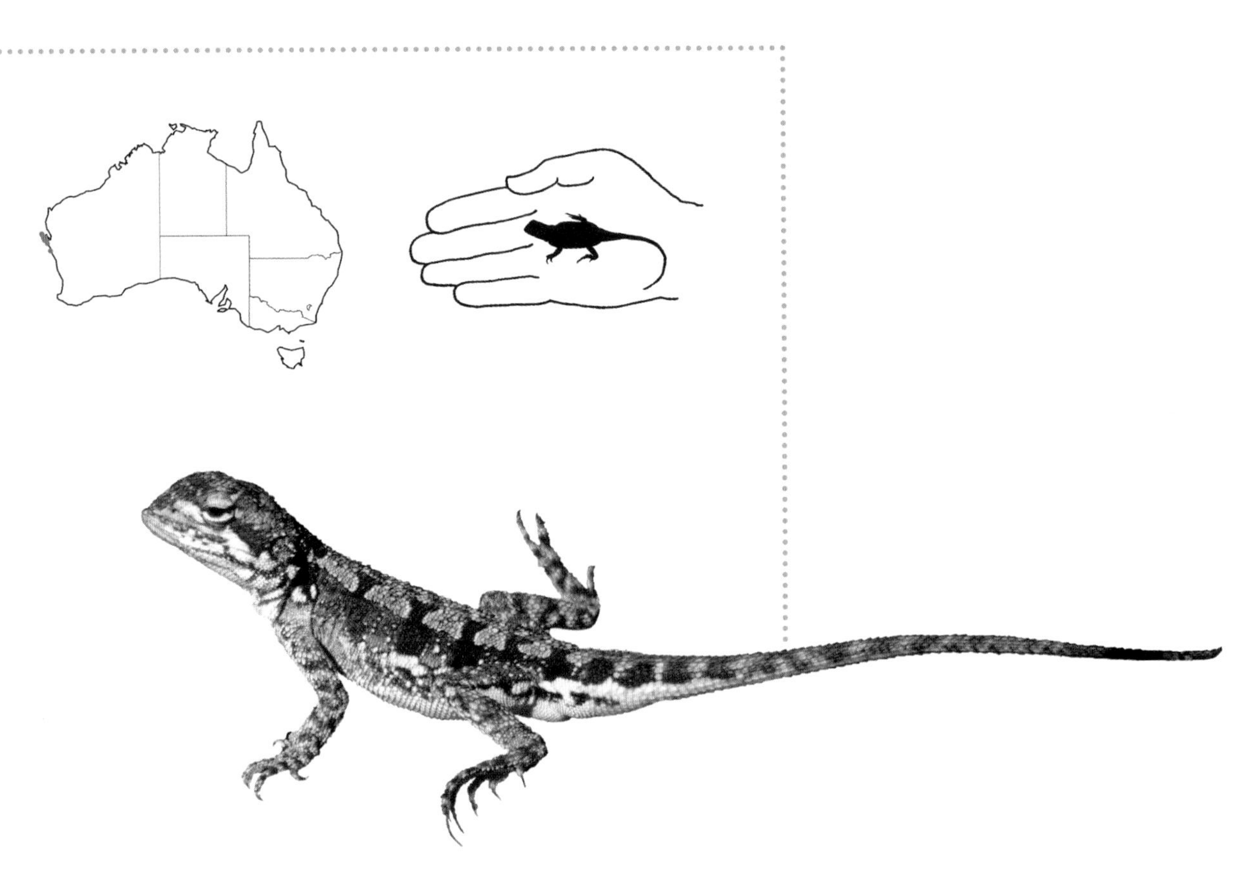

this page
*Ctenophorus butlerorum.*
Tamala Station, WA.

photo S. K. Wilson

opposite page
right:
*Ctenophorus butlerorum.* Tamala Station area, WA.

photo S. K. Wilson

left:
*Ctenophorus butlerorum* habitat. Tamala Station, WA.
photo B. Maryan

## WESTERN RING-TAILED DRAGON

*Ctenophorus caudicinctus* Günther, 1875

**DESCRIPTION:** SVL 90 mm. Head and body relatively deep and robust, with long limbs and tail. Body scales mostly homogeneous; overlapping, with spinose keels converging back towards midline, largest on back and becoming much smaller and smooth to weakly keeled on flanks. Nuchal crest present, comprising small raised laterally compressed scales, and a raised vertebral ridge of aligned keels on slightly enlarged vertebral scales along back. Sexes differ in Pilbara, but less so further south. Mature Pilbara males develop laterally compressed body and tail. The head and back is dull blood red, with pigment tending to concentrate to form about 3 wavy longitudinal reddish streaks on flanks, and the tail is yellowish brown with prominent narrow dark rings. Females and juveniles are reddish brown with 2–4 longitudinal rows of large dark spots alternating with thin pale bands or transverse rows of smaller spots. Individuals from populations along the coast of the Great Sandy Desert, from 80 Mile Beach, south to Pardoo have a distinctive slate-grey background colour on head and body. Femoral and preanal pores 23–40.

**KEY CHARACTERS:** Differs from the Goldfields Ring-tailed Dragon (*C. infans*) in being larger and exhibiting weak to strong sexual dimorphism (versus SVL 67 mm and mature males similar to females and juveniles).

**DISTRIBUTION AND ECOLOGY:** Mainly rock inhabiting on a variety of ranges and outcrops, and extending onto adjacent gibber flats. Northerly populations, along the coastal region of the Great Sandy Desert, inhabit boulder outcrops on sandy soils. Associated with various vegetation types such as *Triodia* grasses and Acacia shrublands and woodlands. Lizards generally select protruding rocks as elevated vantage points and retreat to crevices if threatened.

**BIOLOGY:** Clutches of 3–7 eggs are recorded, and of 68 active individuals measured near Wiluna in WA, average body temperature was 37.2° C. While it consumes a variety of arthropods, termites and ants were significant prey comprising 64% and 20% by volume respectively (Pianka, 2014). Individuals of the Pilbara population hatch between January and May, males begin to exhibit adult characteristics between June and September, maturing between October and March, and have all died by the following June. Females are gravid between November and March, and they too have perished by June (Storr, 1967).

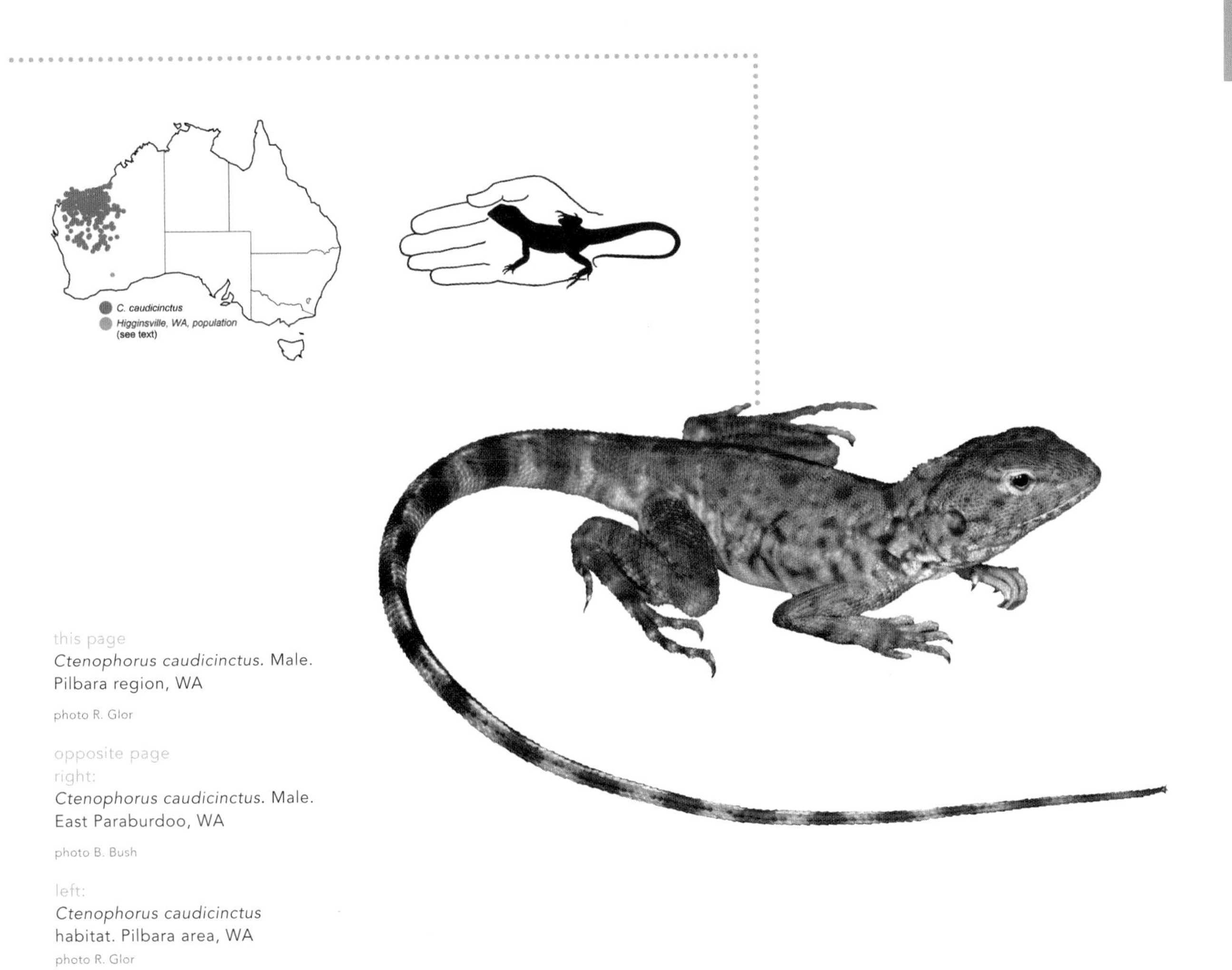


this page
*Ctenophorus caudicinctus*. Male. Pilbara region, WA

photo R. Glor

opposite page
right:
*Ctenophorus caudicinctus*. Male. East Paraburdoo, WA

photo B. Bush

left:
*Ctenophorus caudicinctus* habitat. Pilbara area, WA
photo R. Glor

**NOTES:** Recent genetic studies have revised the concept of *Ctenophorus caudicinctus*, with some subspecies elevated to the status of distinct species and others synonymised (Melville *et al*, 2016). In this new context *Ctenophorus caudicinctus* applies only to the taxa formerly known as *C.c. caudicinctus* and *C. c. mensarum*. An isolated population, from Higginsville, WA, (marked as blue on the distribution map) has been identified morphologically as *C. caudicinctus*. However, genetic research has documented hybrid individuals between *C. caudicinctus* and *C. ornatus* at this location (Melville *et al*, 2017d).

*Ctenophorus caudicinctus.*
Cue area, WA
photo B. Bush

*Ctenophorus caudicinctus.*
Sandfire, WA
photo M. Bruton

# CTENOPHORUS

## SOUTHERN HEATH DRAGON

*Ctenophorus chapmani* Storr, 1977

**DESCRIPTION:** SVL 52 mm. Very small with short limbs and tail. Body scales heterogeneous; one or more erect, multi-keeled tubercles on each side of neck, dorsal body scales variable including keeled and granular scales, numerous erect tubercles on back and flanks particularly within areas of dark colour, and a row of enlarged spinose scales along each side of tail-base. Chin 'terraced' (a straight line forming angular junction along ventrolateral edge of jaw). Grey with a broad vertebral stripe margined by black triangular blotches, 2 narrow dark dorsolateral lines and a row of weak upper lateral blotches. Ventral surfaces of males with intense black reticulations. Femoral and preanal pores 16–25 extending along about three quarters the length of the thigh.

**KEY CHARACTERS:** Differs from other *Ctenophorus* species in this area in having numerous erect tubercules on back. Potentially overlaps with the Nullarbor Earless Dragon (*Tympanocryptis houstoni*) but differs in having an exposed tympanum and more femoral and preanal pores.

**DISTRIBUTION AND ECOLOGY:** Mallee woodlands and heaths on sandy soils across southern Australia, between the Stirling Ranges, WA and Yorke Peninsula, SA. On the Nullarbor Plain it occupies chenopod shrublands. Fully terrestrial, rarely utilising elevated perches. Not particularly swift when compared to other small dragons, tending to rely mainly on its cryptic disruptive pattern to avoid predators. When pursued it scuttles rather than sprints to the cover of low vegetation

**BIOLOGY:** Ants probably form a major component of the diet.

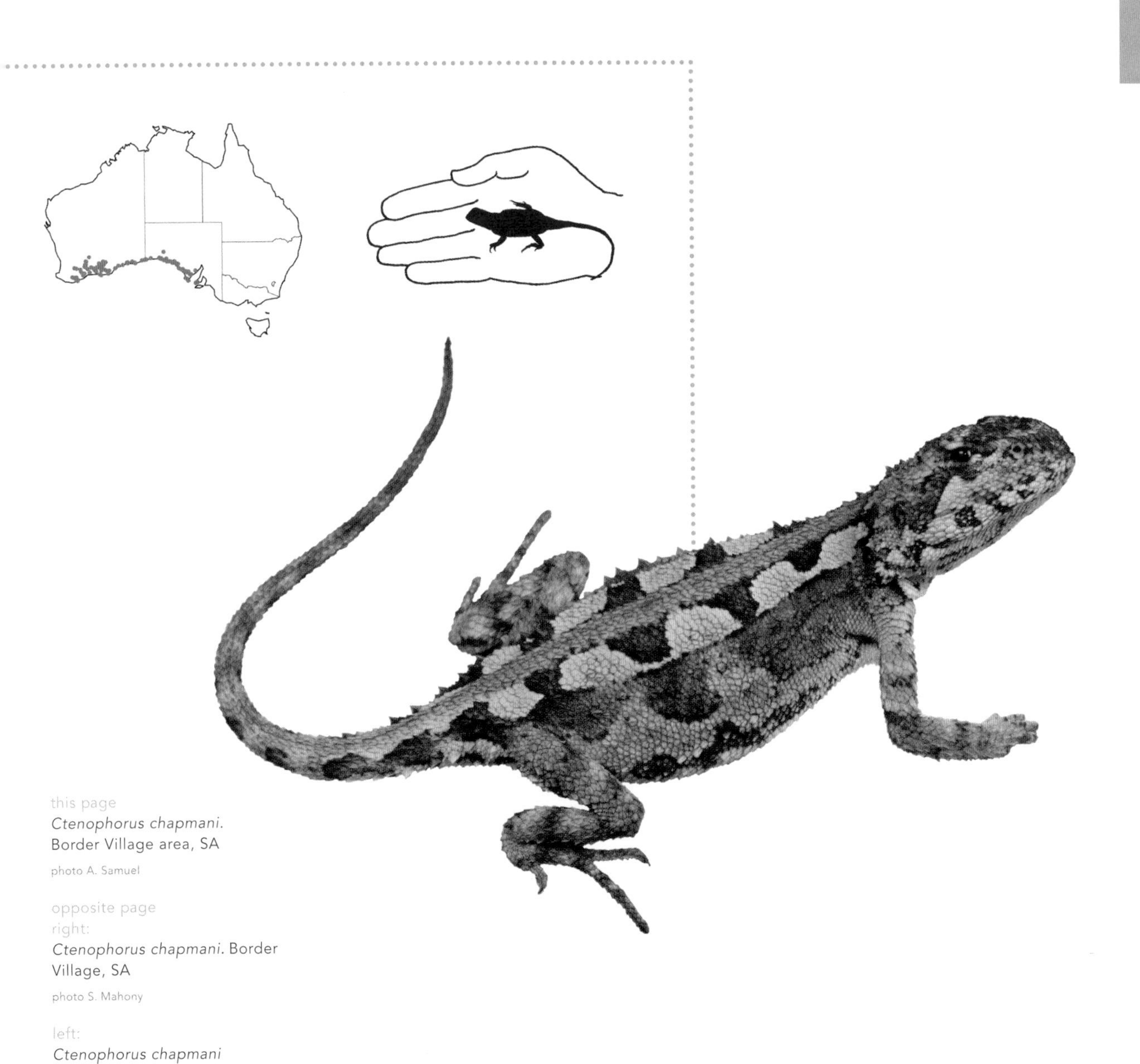

this page
*Ctenophorus chapmani*. Border Village area, SA
photo A. Samuel

opposite page
right:
*Ctenophorus chapmani*. Border Village, SA
photo S. Mahony

left:
*Ctenophorus chapmani* habitat. Point Brown, SA
photo S. K. Wilson

### BLACK-COLLARED DRAGON

*Ctenophorus clayi* Storr, 1966

**DESCRIPTION:** SVL 58 mm. Robust with a short, deep head and relatively short limbs and tail. Scales mostly homogeneous: small and smooth with no nuchal or vertebral crests, a prominent conical tubercle on each jowl and sometimes several others on sides of neck and on nape. Fawn to pale reddish brown with a dark reticulum, a narrow pale vertebral stripe and a series of irregular dark blotches and/or dark-edged pale blotches. Prominent black patch present on side of neck on gular fold, sometimes extending around throat to form a collar. Mature males have prominent yellow patch on either side of gular fold. Femoral and preanal pores 4–9.

**KEY CHARACTERS:** Differs from all other sympatric dragons in possessing black neck patches and collar along gular fold.

**DISTRIBUTION AND ECOLOGY:** Red sand ridges and sand plains vegetated with spinifex in central and western deserts. A possibly isolated population occurs on North West Cape, WA. It shelters in short burrows dug under the bases of spinifex hummocks. *C. clayi* is a habitat generalist (Pianka, 2013a) and has been found to be in highest abundance in habitats with low rainfall (Pastro *et al*, 2013).

**BIOLOGY:** Perches on shrubs, tussocks and woody debris, dashing on all fours to its burrow or under vegetation if approached. *C. clayi* actively thermoregulates, with a preferred body temperature of 36.9°C (Pianka, 2013a). In the Great Victoria Desert the diet of this species is dominated by grasshoppers (30.4%), with termites and ants comprising a much lesser proportion of their diet (Pianka, 2013a). Clutches of 2–4 eggs have been recorded (Greer, 1989).

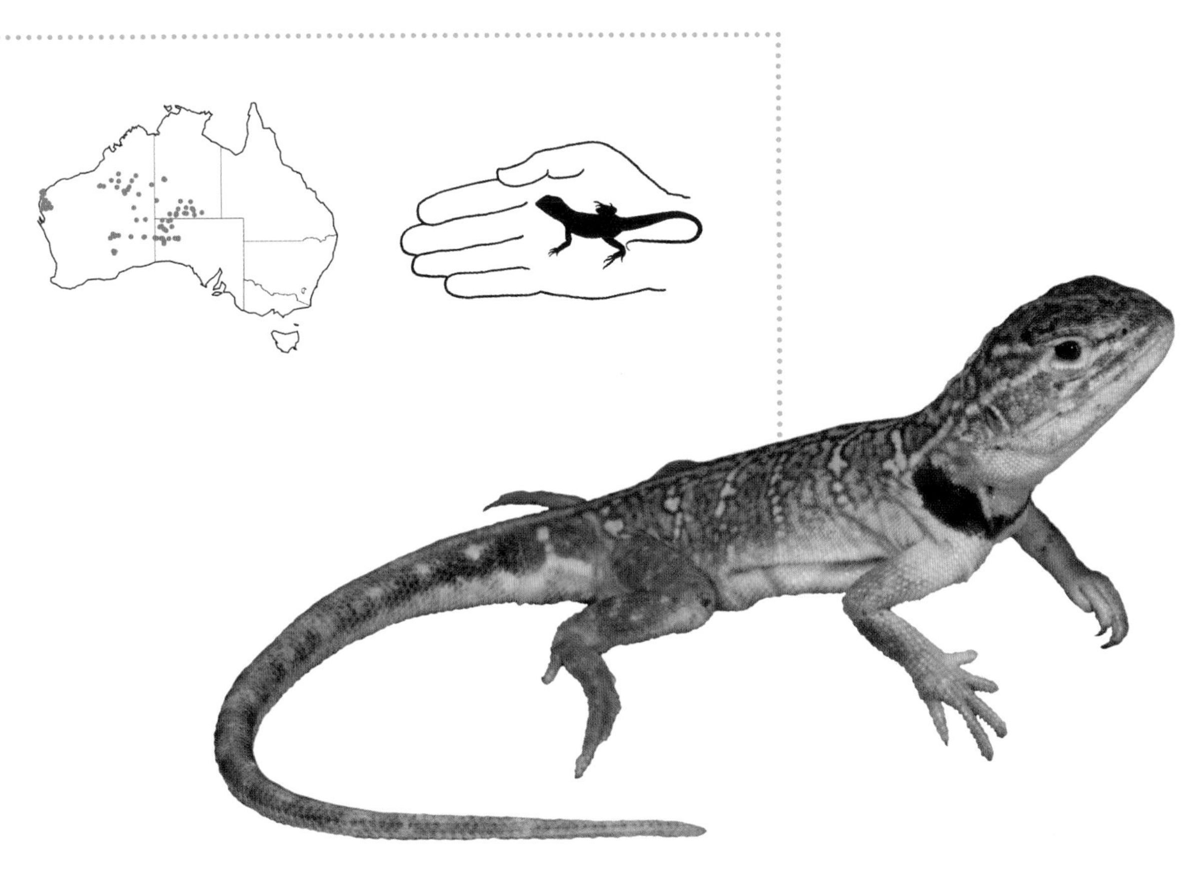

this page
*Ctenophorus clayi*.
Learmonth area, WA

photo A. Elliott

opposite page
right:
*Ctenophorus clayi*.
Condun Well, WA
photo S. K. Wilson

left:
*Ctenophorus clayi* habitat.
Bullara Station, WA
photo S. K. Wilson

## CRESTED DRAGON; BICYCLE LIZARD

*Ctenophorus cristatus* Gray, 1841

**DESCRIPTION:** SVL 110 mm. Head moderately long and narrow, body slender and limbs and tail very long. Body scales small and uniform with prominent nuchal crest of compressed triangular scales and dorsolateral folds with spinose scales, decreasing in size posteriorly. Colour and pattern brightest on mature males but capable of extreme colour change, becoming drab when cold, inactive or stressed. Head, forelimbs and anterior body cream, yellow to rich reddish brown with prominent black reticulum and broken black dorsolateral streak to midbody or hips. Posterior body, hind limbs and base of tail greyish brown and remaining three quarters of tail prominently ringed with black and cream to pale orange. Ventral markings are black on males and grey on females; a broad streak along centre of throat edged with wavy lines, a black triangular patch on chest extending onto forelimbs and back to abdomen and a broad band over vent onto hind limbs. A total of 42–64 preanal and femoral pores extend in a line along full length of thigh.

**KEY CHARACTERS:** Differs from the Lozenge-marked Dragon (*C. scutulatus*) and McKenzie's Dragon (*C. mckenziei*) in having dorsolateral crests.

**DISTRIBUTION AND ECOLOGY:** Semi-arid woodlands across southern Australia, particularly mallee but also salmon gums, mulga and occasionally occurs in areas of saltbush. Has not been found to occupy agricultural or highly modified habitats (Williams *et al*, 2012). Although this species has been reported to use burrows (Greer, 1989), more recent work has suggested that, at least in WA, *C. cristatus* uses hollow logs as retreats or sleeps in the open (Thompson & Withers, 2005). Extremely swift and terrestrial, perching on the ground or low timber and dashing on hind limbs with forelimbs held to chest if disturbed. The bipedal gait has led to the name Bicycle Lizard.

**BIOLOGY:** Clutches of 2–9 eggs are recorded (Houston & Hutchinson, 1998). A generalist arthropod predator, but ants and grasshoppers are reported as significant food items, and small lizards are also taken (Pianka, 1971b).

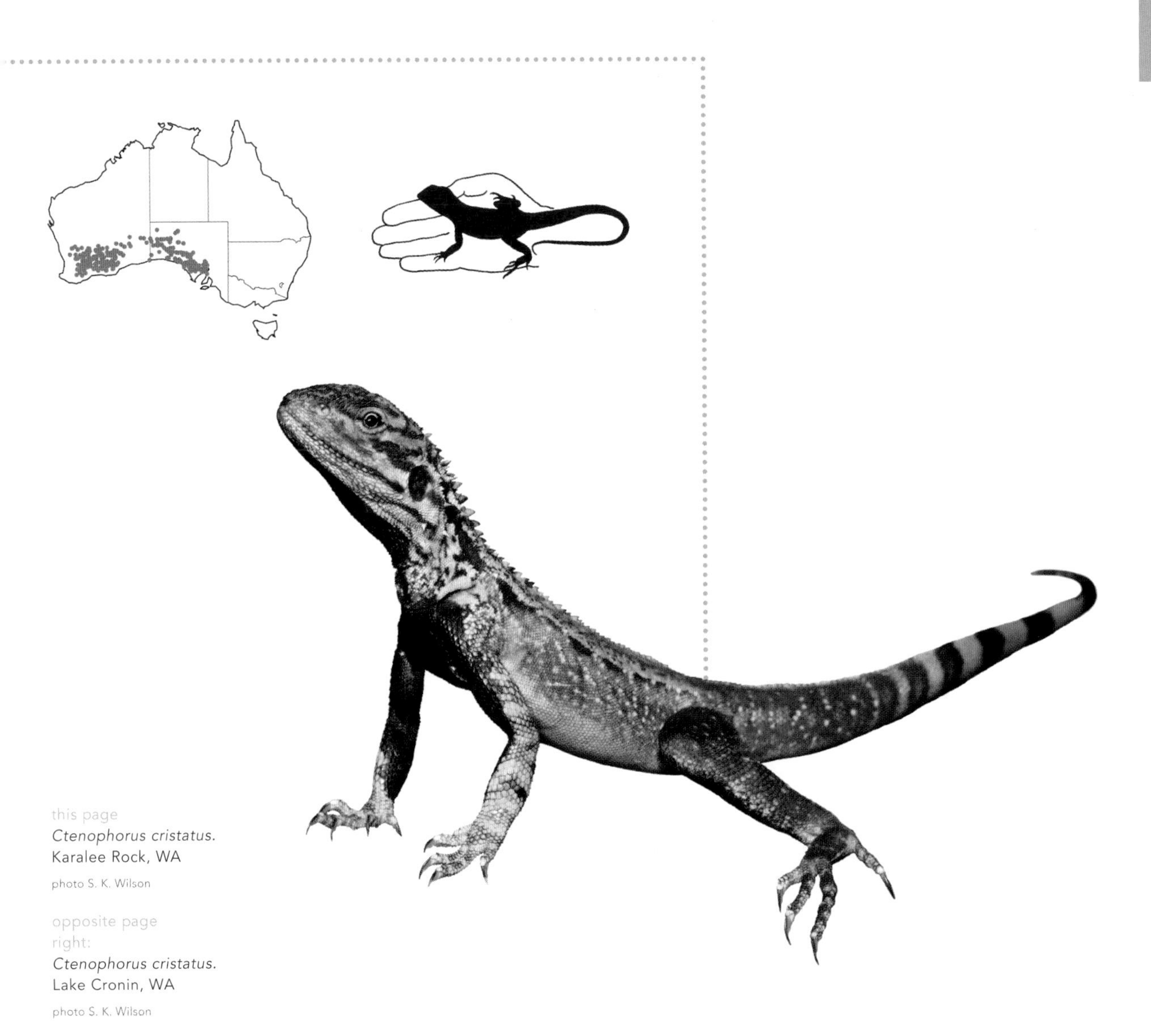

this page
*Ctenophorus cristatus.*
Karalee Rock, WA

photo S. K. Wilson

opposite page
right:
*Ctenophorus cristatus.*
Lake Cronin, WA

photo S. K. Wilson

left:
*Ctenophorus cristatus*
habitat. Lake Hurlstone, WA
photo S. K. Wilson

*Ctenophorus cristatus.*
Goongarrie Station, WA
photo J. Melville

## TAWNY DRAGON

*Ctenophorus decresii* Duméril and Bibron, 1837

**DESCRIPTION:** SVL 82 mm. Head and body dorsally depressed. Scales on snout smooth to keeled, and scalation on body mostly homogeneous; small uniform scales, each with a low keel, and a low crest of enlarged, strongly keeled scales on nape. Southern population, from Kangaroo Island and the southern Mt Lofty Ranges, has small, mostly pale tubercles scattered on flanks. Colouration differs markedly between the sexes, and between northern and southern males. Males are bluish grey to brown with a broad darker upper lateral stripe. On southern males, this is constricted into blotches. It is continuous on northern males, from the northern Mt Lofty and Flinders Ranges. The anterior upper margin is white, yellow, orange to red. Southern males have prominent blue throats with yellow to orange along their gular folds, while northern males are more variable featuring orange, yellow, orange and yellow, or grey throats. Females and juveniles are brown to grey, darker on flanks, with scattered dark flecks. Femoral pores 34–45 (northern males); 40–46 (southern males).

**KEY CHARACTERS:** Differs from the Red-barred Dragon (*C. vadnappa*) and the Ochre Dragon (*C. tjantjalka*) in having smooth to keeled scales on the snout (versus wrinkled) and no vertebral keel line on back. Southern population differs from the Peninsula Dragon (*C. fionni*) in having scattered pale tubercles on flanks (versus absent). Differs from the Barrier Range Dragon (*C. mirrityana*) in lacking a dark median throat stripe on males.

**DISTRIBUTION AND ECOLOGY:** Rock inhabiting, on outcrops and ranges from Kangaroo Island to the Mt Lofty and Flinders Ranges and Olary Spur, SA. Tawny Dragons occupy habitats ranging from well-watered to semi-arid. While adults occur almost exclusively on rocks, juveniles are often encountered among surrounding vegetation. They perch on elevated sites, and if disturbed dash beneath a rock slab or into a crevice.

**BIOLOGY:** Clutches of 3–7 eggs are recorded (Greer, 1989). Males are territorial and perform complex and impressive displays to potential rivals. They align themselves parallel to each other, often facing opposite directions, lower their brightly coloured throats, raise their nuchal crests and vertebral ridges, elevate and lower their bodies by raising their hind limbs, and coil their tails horizontally away from their opponents (Gibbons, 1979). Social behaviour, and the evolutionary and genetic basis colour polymorphisms have been well studied in this species (e.g., Stuart-Fox *et al*, 2003; McLean *et al*, 2015; McLean *et al*, 2017).

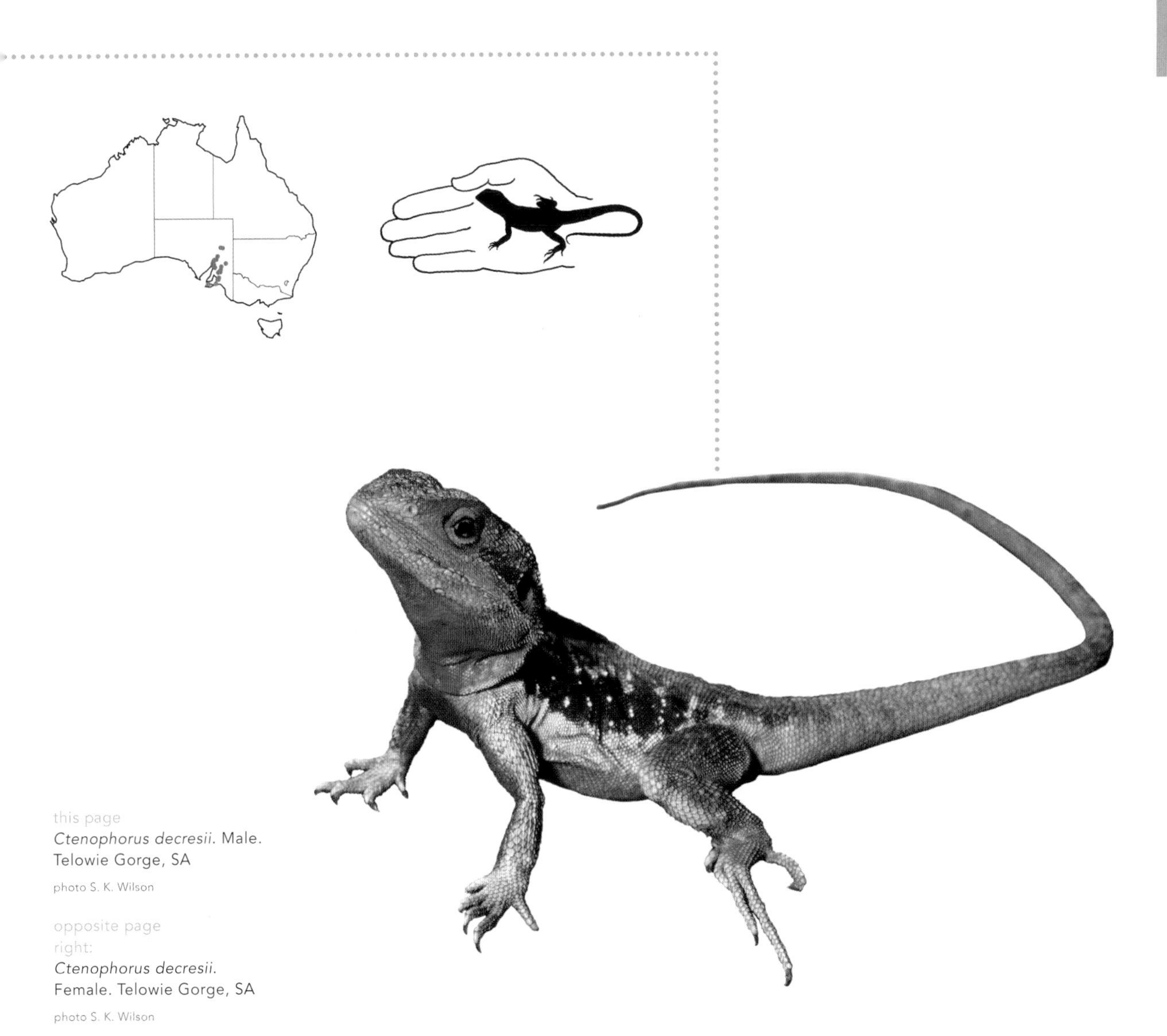

this page
*Ctenophorus decresii*. Male. Telowie Gorge, SA

photo S. K. Wilson

opposite page
right:
*Ctenophorus decresii*. Female. Telowie Gorge, SA

photo S. K. Wilson

left:
*Ctenophorus decresii* habitat. Wilpena Pound area, SA

photo S. K. Wilson

*Ctenophorus decresii*. Male.
Telowie Gorge, SA
photo S. K. Wilson

*Ctenophorus decresii*. Male.
Wilpena Pound area, SA
photo S. K. Wilson

## LONG-TAILED SAND DRAGON

*Ctenophorus femoralis* Storr, 1965

**DESCRIPTION:** SVL 57 mm. Slender, with weakly depressed head and body, long limbs and a very long slender tail. Body scales mostly homogeneous; small uniform dorsal scales, each with a keel converging back towards midline. Scales on flanks smaller and smooth to weakly keeled. There are no crests, spines or enlarged scales. Brick red to dull orange with weak pattern comprising small pale spots, smaller dark flecks, usually a pair of obscure pale broken dorsolateral stripes and a more prominent pale midlateral stripe. Mature males have a black patch on chest. A total of 18–32 femoral and preanal pores confined to inner halves of thighs and curving slightly forward towards midline.

**KEY CHARACTERS:** Differs from the Central Military Dragon (*C. isolepis*), the Rufus Military Dragon (*C. rubens*) and the Spotted Sand Dragon (*C. maculatus*) in having pores confined to inner halves of thighs (versus extending full length of thighs). Differs further from *C. isolepis* and *C. rubens* by having preanal pores curving slightly forwards at midline (versus curving sharply forward).

**DISTRIBUTION AND ECOLOGY:** Red sand ridges and adjacent sand plains vegetated with spinifex on arid midwest coast and hinterland of WA, between North West Cape and Kennedy Range.

**BIOLOGY:** Extremely swift and wholly terrestrial, foraging in open spaces and around the edges of low vegetation, mainly on crests and slopes of dunes, and dashing on all four limbs ahead of any potential approaching danger. It does not utilise any elevated perches, and shelters under vegetation rather than in burrows. Diet probably comprises almost exclusively of ants. Lifespan is probably annual.

**NOTES:** Research has shown that this species belongs to the sand dragon species complex, which also includes *C. maculatus* and *C. fordi* (Edwards *et al*, 2015). These sand dragons consist of 11 lineages occupying a range of sandy habitats. Taxonomic work is underway.

opposite page left:
*Ctenophorus femoralis* habitat.
Bullara Station, WA
photo S. K. Wilson

opposite page right:
*Ctenophorus femoralis*.
Bullara Station, WA
photo S. K. Wilson

below:
*Ctenophorus femoralis*.
Bullara Station, WA
photo S. K. Wilson

## PENINSULA DRAGON

*Ctenophorus fionni* Procter, 1923

**DESCRIPTION:** SVL 96 mm. Head and body dorsally depressed. Snout scales smooth to keeled, and body scalation mostly homogeneous; small and uniform, with a low crest of enlarged, strongly keeled scales on nape. Colouration differs markedly between the sexes. Males exhibit extensive geographic variation and at least six colour variants have been recognised (Houston & Hutchinson, 1998).

- Northern males (Port Augusta north to Woomera and Andamooka) are brown grading to blackish-brown on neck, shoulders and flanks, with pale blotches, becoming reddish and tending to form vertical bars on flanks.
- Central males (Gawler Ranges and central to northern Eyre Peninsula) have a black back with small round spots merging to orange on the shoulders, nape and neck.
- Port Lincoln males (southern Eyre Peninsula) are brown with heavy black speckling. Numerous white spots, some black-edged, are scattered over back and flanks.
- Southern males (from Jussieu Peninsula to Mt Wedge) are pale grey, darker on neck and anterior flanks, with pale markings reduced to creamish spots on skin folds along sides of neck and white spots along dorsolateral folds on body.
- Western males (coast and islands west of Cape Finniss) resemble southern males, but with black along flanks and sides of neck, prominently edged by pale spots,
- Neptune and Wedge Island males are brown, darkening to black on neck, shoulders and flanks, with prominent dark-edged white spots across back.

Females and juveniles are brown to brick red with sparse to dense darker mottling, darker on flanks. Femoral and preanal pores 32–47.

**KEY CHARACTERS:** Differs from the southern population of Tawny Dragon (*C. decresii*) in lacking scattered pale tubercles on flanks (versus present). Difficult to distinguish from the northern population except by male colouration.

**DISTRIBUTION AND ECOLOGY:** Rock inhabiting, on outcrops and ranges of Eyre Peninsula and off-shore islands, extending north to Tarcoola and the western edge of Lake Torrens, SA. It occupies a variety of rock types in habitats ranging from well-watered to semi-arid. The restriction to rocky habitats has probably given rise to the distinctive, colour forms, each isolated from its neighbour by tracts of unsuitable terrain. Lizards perch on elevated sites, and if disturbed take refuge beneath a rock slab or in a crevice.

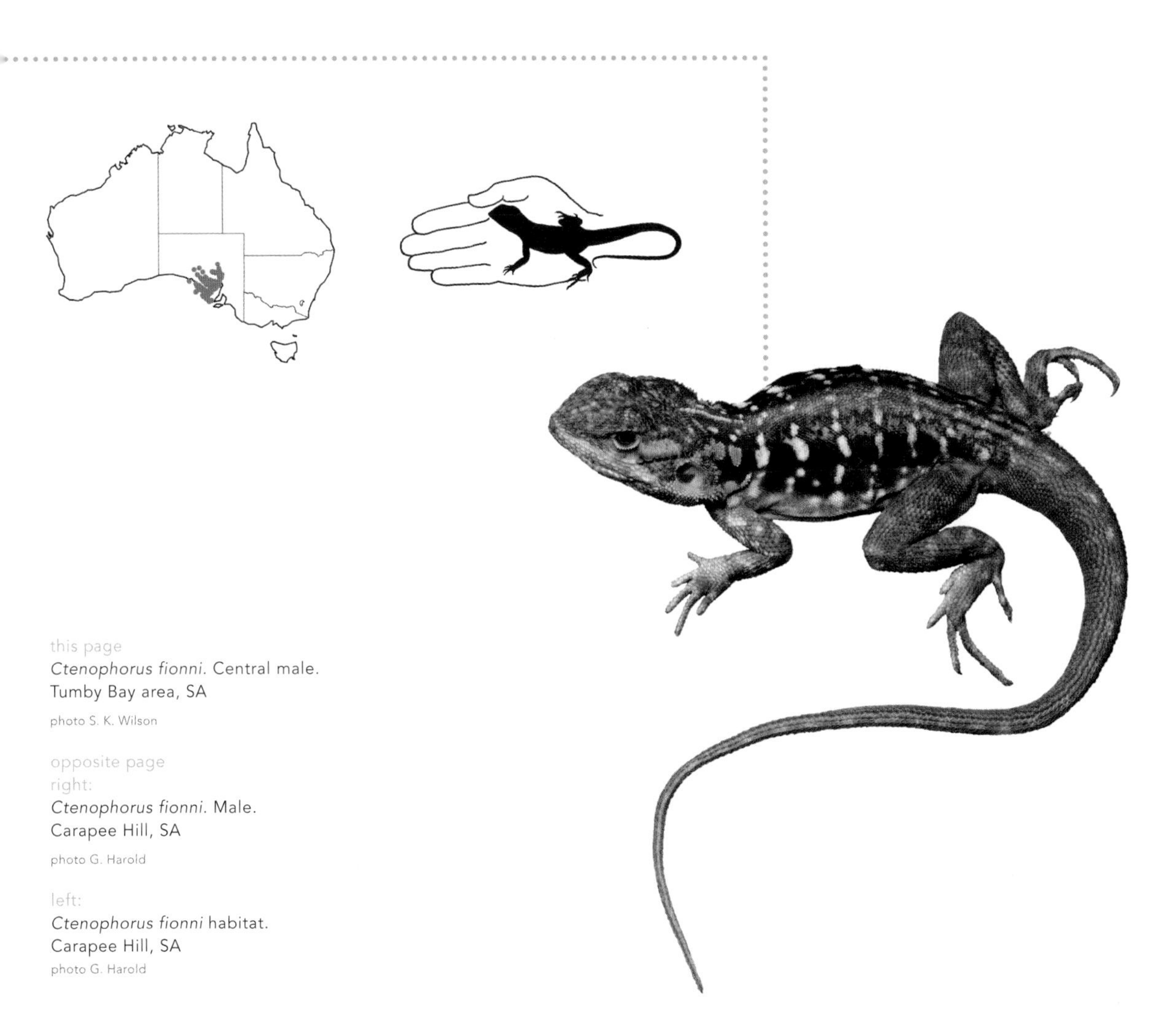

this page
*Ctenophorus fionni*. Central male.
Tumby Bay area, SA
photo S. K. Wilson

opposite page
right:
*Ctenophorus fionni*. Male.
Carapee Hill, SA
photo G. Harold

left:
*Ctenophorus fionni* habitat.
Carapee Hill, SA
photo G. Harold

**BIOLOGY:** Territorial males perform to potential rivals by aligning themselves parallel to each other, often facing opposite directions. They lower their brightly coloured throats, raise their nuchal crests and vertebral ridges, elevate and lower their bodies by raising their hind limbs, and coil their tails horizontally or obliquely away from their opponents (Gibbons, 1979). As with many dragons, adult males are larger than females but have been shown to have a higher mortality rate than females, possibly due to predation (Johnston, 2011). Clutch sizes of 2–6 eggs (Johnson, 1999), with most females only producing one clutch each season but a small percent will lay a second clutch.

top:
*Ctenophorus fionni*. Male.
NE of Dutton Bluff, SA
photo G. Harold

bottom:
*Ctenophorus fionni*. Female.
Carapee Hill, SA
photo G. Harold

opposite page:
*Ctenophorus fionni*.
Western male.
Calca area, SA
photo S. K. Wilson

## MALLEE SAND DRAGON

*Ctenophorus fordi* Storr, 1965

**DESCRIPTION:** SVL 58 mm. Head and body weakly depressed, with long limbs and a very long slender tail. Body scales mostly homogeneous; small uniform dorsal scales, each with a keel converging back towards midline. Scales on flanks smaller and smooth to weakly keeled. There are no crests, spines or enlarged scales. Brown to reddish brown with prominent pale dorsolateral and lateral stripes edged above, between and below with longitudinal series of black blotches. Males have a black chest, and throats marked with black spots or bars, coalescing on western populations to form a V-shape. A total of 22–43 femoral and preanal pores extend in a straight line up to three quarters of the way along thighs and curve slightly forwards at midline.

**KEY CHARACTERS:** Differs from the Central Military Dragon (*C. isolepis*) and the Spotted Sand Dragon (*C. maculatus*) in having femoral pores not extending beyond three quarters of thigh (versus extending full length of thigh). Differs further from *C. isolepis* in having preanal pores curving slightly forwards towards midline (versus curving sharply forward).

**DISTRIBUTION AND ECOLOGY:** Sandy areas across the mallee belt of arid to semiarid southern Australia. Mainly associated with spinifex, usually in association with mallee eucalypts, but in some areas of SA it occupies dunes vegetated with cane grass (*Zygochloa*) or blue bush (*Maireana*) (Houston & Hutchinson, 1998). Populations are fragmented by areas of unsuitable habitat.

**BIOLOGY:** Extremely swift and wholly terrestrial, foraging in open spaces and around the margins of low vegetation. It dashes on all four limbs ahead of any potential danger, stopping frequently to assess the approaching threat. It does not utilise any elevated perches, and shelters under vegetation rather than in burrows. Diet consists almost exclusively of ants but other insects such as flies are also eaten. Males emerge from winter inactivity in August, 4–5 weeks before the females. Between October and January females display to males by arching their hindquarters and tail off the ground, though this is not believed to be associated with mating (Cogger, 1978). Females have not been found to choose any particular male over another (Olsson, 2001a). *C. fordi* has been the focus of research on reproductive behaviour and competition between males (e.g., Olsson, 2001b; Uller *et al*, 2013). Differences in body patterning between the sexes has been found to be a trade-off between attracting a mate and camouflage (Edwards *et al*, 2015). *C. fordi* has also been found to have ultraviolet body patterning that varies significantly between males and females (Garcia *et al*, 2013). Clutches of 2–3 eggs are laid, and up to 3 clutches may be produced per season. Lifespan is mainly annual with some lizards living up to 2 years (Cogger, 1978).

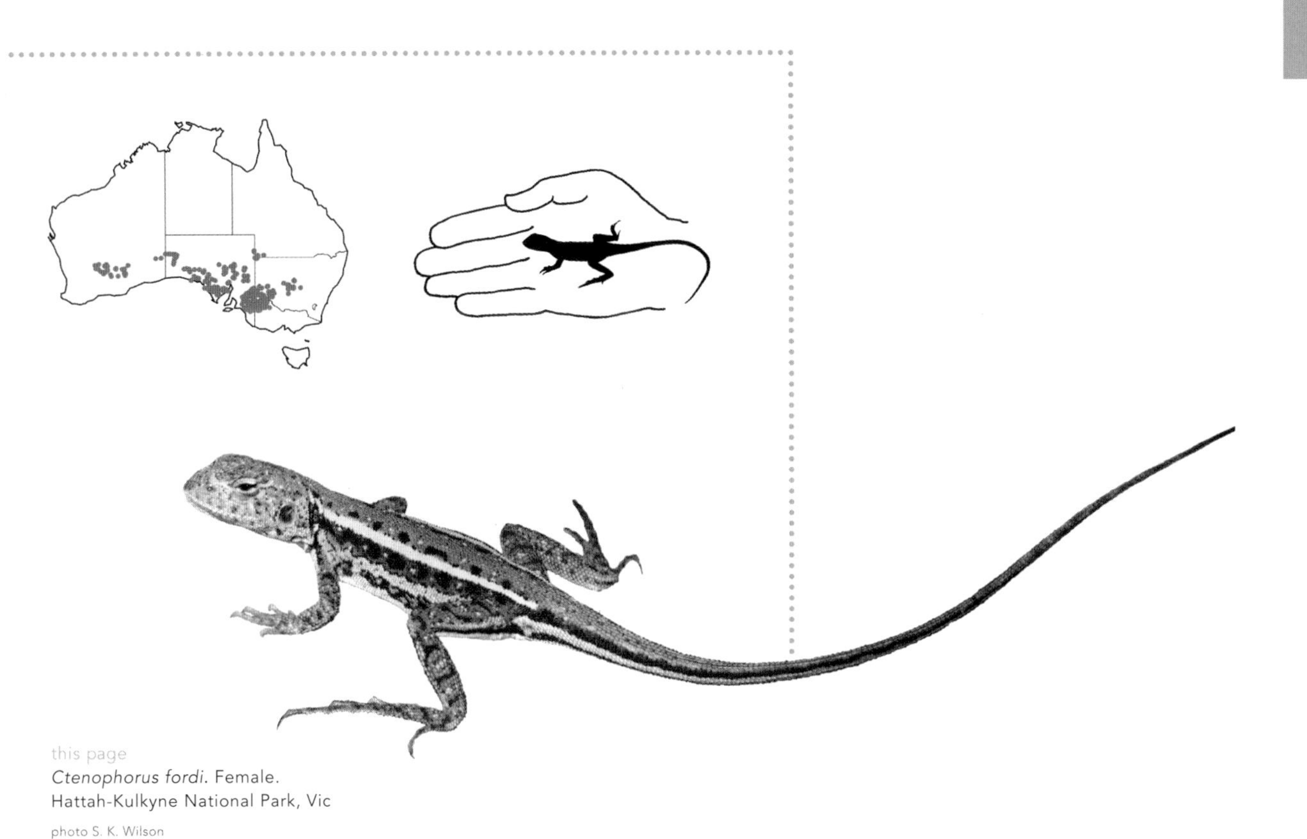

this page
*Ctenophorus fordi*. Female.
Hattah-Kulkyne National Park, Vic
photo S. K. Wilson

opposite page
right:
*Ctenophorus fordi*. Male.
Menzies area, WA
photo S. K. Wilson

left:
*Ctenophorus fordi* habitat.
Hattah-Kulkyne National Park, Vic
photo S. K. Wilson

**NOTES:** *C. fordi* belongs to the sand dragons, a group closely allied to the military dragons (*C. isolepis* and *C. rubens*). These sand dragons, which also include *Ctenophorus maculatus* and *Ctenophorus femoralis*, consist of 11 lineages occupying a range of sandy habitats (Edwards *et al*, 2015). Taxonomic work is underway.

*Ctenophorus fordi*. Male.
Hattah-Kulkyne National Park, Vic
photo S. K. Wilson

## GIBBER DRAGON

*Ctenophorus gibba* Houston, 1974

**DESCRIPTION:** SVL 82 mm. Robust with a short, deep head and relatively short limbs and tail. Scales mostly homogeneous: small and convex with several spinose tubercles on the back of the head and along the nape. Yellowish brown, grey to reddish brown with a fine dark reticulum or flecks, 6–8 pairs of dark spots between nape and hips and often a similar or more fragmented series on flanks. Sides of tail marked with 20–30 dark squarish blotches. Ventral pattern is very distinctive: an elongate black blotch on the chin, a larger round blotch on the throat, and a large patch on the chest. A pale yellow suffusion may be present on anterior chest and shoulders. Femoral and preanal pores 26–35.

**KEY CHARACTERS:** Differs from the Central Netted Dragon (*C. nuchalis*) and the Western Netted Dragon (*C. reticulatus*) in having an elongate black blotch on the chin and a round blotch on the throat of males.

**DISTRIBUTION AND ECOLOGY:** Restricted to the arid north-eastern interior of SA, from around Maree to Dalhousie Springs. It occurs on sparsely vegetated plains of soft loamy soils strewn with gibber stones. It basks on low protruding rocks and shelters in oblique burrows 20–50 centimetres long, in soft soil between stones.

**BIOLOGY:** Because they live in an environment with high summer temperatures, Gibber Dragons remain inactive in their burrows during the hottest periods. The same applies to winter when temperatures drop significantly. However, when active they are conspicuous as they perch on stones against a featureless backdrop. When approached they have been recorded to dive into soft, floury clay much like some sand-swimming skinks.

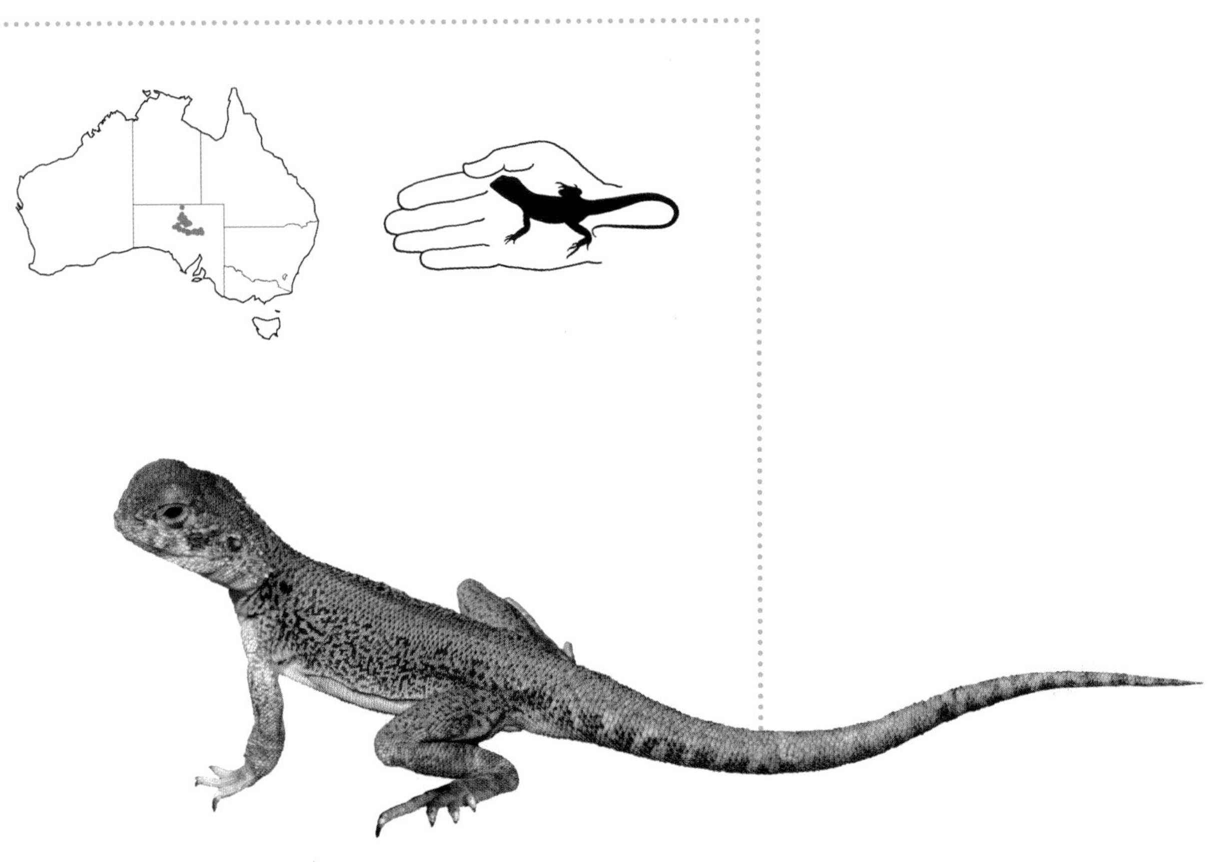

this page
*Ctenophorus gibba.*
Coober Pedy—William Creek Rd, SA

photo S. K. Wilson

opposite page
right:
*Ctenophorus gibba.*
Coober Pedy—William Creek Rd, SA

photo S. K. Wilson

left:
*Ctenophorus gibba* habitat. Marree area, SA
photo S. K. Wilson

## GRAAF'S RING-TAILED DRAGON; TANTALKA

*Ctenophorus graafi* Storr, 1967

**DESCRIPTION:** SVL 79 mm. Head and body relatively deep and robust, with long limbs and tail. Body scales mostly homogeneous; overlapping with weak to moderately strong spinose keels converging back towards midline, largest on back becoming much smaller and smooth to weakly keeled on flanks. Nuchal crest present, comprising small raised laterally compressed scales, and a raised vertebral ridge of aligned keels on slightly enlarged vertebral scales. Sexes differ. Males are reddish brown on back, merging to grey on flanks, with obscure dark bands on distal three quarters of tail. Small individuals have row of dark paravertebral spots, but little or no indication of the transverse pale spots present on related ring-tailed dragon species. There is a large dark chest patch, and a salmon pink suffusion around the gular fold. Females are brighter reddish brown with dark flecks and paravertebral series of dark spots which are poorly aligned transversely. Femoral and preanal pores 29–39.

**KEY CHARACTERS:** Differs from the nearest populations of Slater's Ring-tailed Dragon (*C. slateri*) in having dorsal keels weak to moderately strong, the same colour as dorsal scales (versus dorsal keels strong and black). Differs from nearby Rusty Dragon (*C. rufescens*) in having a deeper head and body (versus strongly flattened) and fewer femoral and preanal pores.

**DISTRIBUTION AND ECOLOGY:** Rock inhabiting, on rocky ranges and granite outcrops along the northern edge of the Great Victoria Desert, from Warburton Range, east to the Barrow Range, WA. Lizards generally select protruding rocks as elevated vantage points and retreat to crevices if threatened.

**BIOLOGY:** Poorly known but believed to be broadly similar to its close relative, the Western Ring-tailed Dragon (*C. caudicinctus*).

**NOTES:** Recent genetic studies (Melville *et al*, 2016) have elevated this species from its previous placement as a subspecies of *Ctenophorus caudicinctus*.

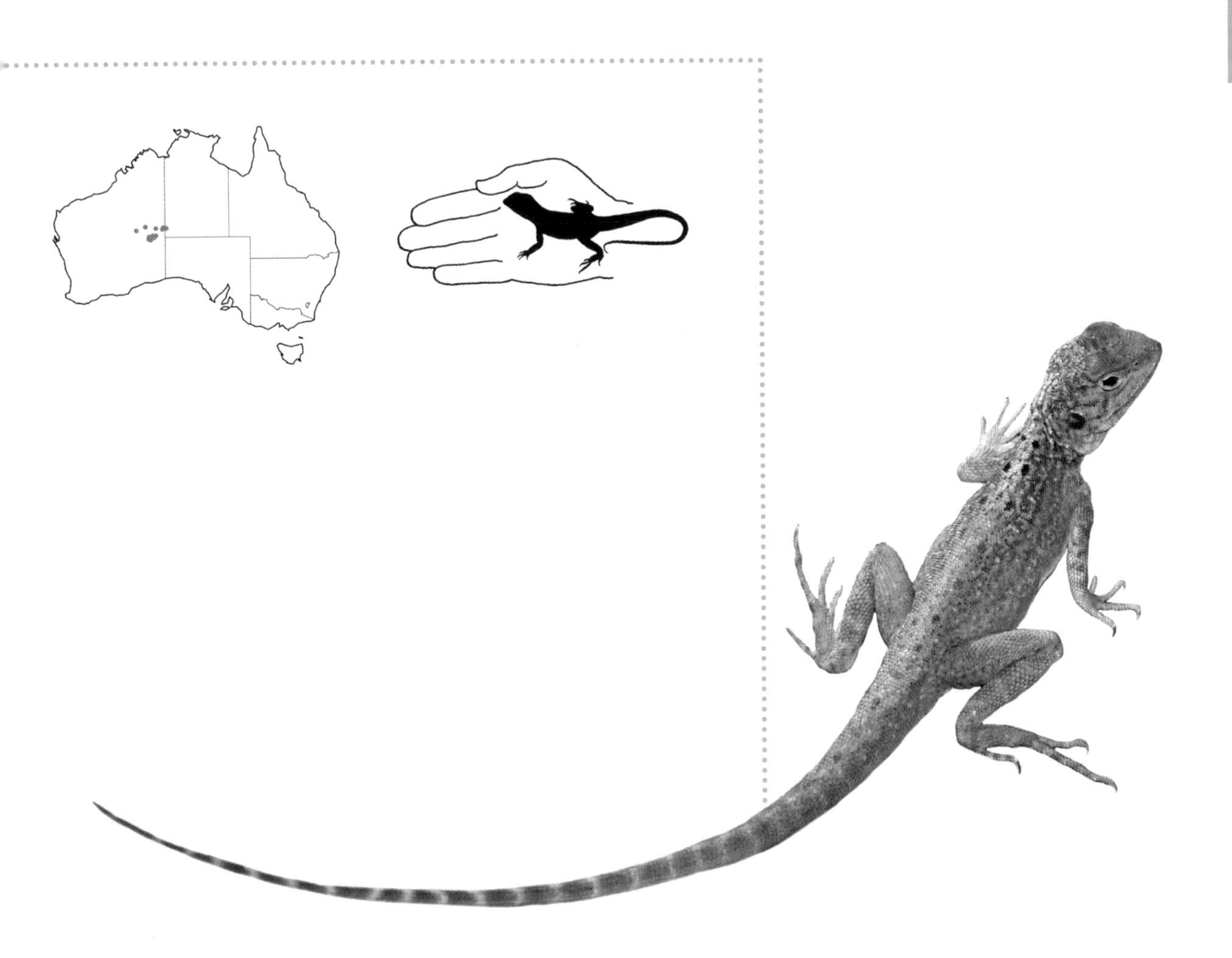

*Ctenophorus graafi.*
Walter James Range, WA
photos M Hutchinson

opposite page:
*Ctenophorus graafi*
habitat. Punkulpirri
Rockhole, Walter James
Range, WA
photo M Hutchinson

## GOLDFIELDS RING-TAILED DRAGON

*Ctenophorus infans* Storr, 1967

**DESCRIPTION:** SVL 67 mm. Head and body relatively deep and robust, with moderate limbs and tail. Body scales mostly homogeneous; overlapping with spinose keels converging back towards midline. Weak nuchal crest present, comprising small, raised laterally compressed scales, and a barely discernible vertebral ridge of slightly enlarged scales. Juveniles and adults of both sexes are similar. Brick red to orange brown with paravertebral series of dark brown dots merging to form bands on tail, and alternating on body with narrow transverse rows of pale dots or lines. Femoral and preanal pores 32–42.

**KEY CHARACTERS:** Differs from the Western Ring-tailed Dragon (*C. caudicinctus*) in being smaller, with little difference between juveniles and adults of both sexes, and relatively shorter appendages (versus SVL 90 mm and weak to strong sexual dimorphism between adults).

**DISTRIBUTION AND ECOLOGY:** Rock inhabiting in the WA Goldfields from Cashmere Downs Station and Laverton, south to Yundamindra Station. Lizards generally select protruding rocks as elevated vantage points and retreat to crevices if threatened.

**BIOLOGY:** Poorly known but believed to be broadly similar to its close relative, the Western Ring-tailed Dragon (*C. caudicinctus*).

**NOTES:** Recent genetic studies (Melville *et al*, 2016) have elevated this species from its previous placement as a subspecies of *Ctenophorus caudicinctus*.

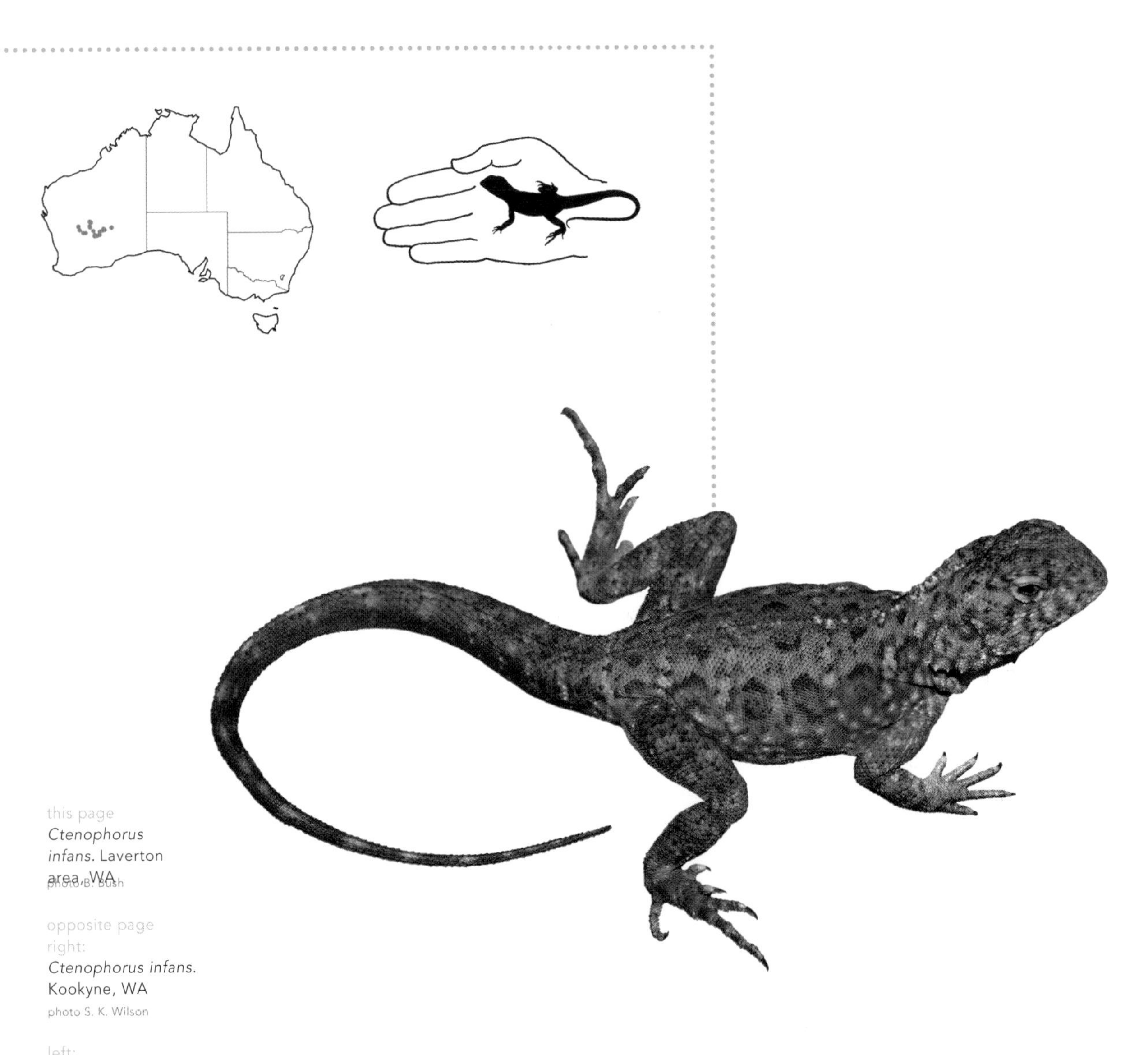

this page
*Ctenophorus infans*. Laverton area, WA
photo B. Bush

opposite page
right:
*Ctenophorus infans*.
Kookyne, WA
photo S. K. Wilson

left:
*Ctenophorus infans* habitat.
Niagara Dam area, WA
photo B. Bush

## CENTRAL MILITARY DRAGON

*Ctenophorus isolepis* Fischer, 1881

**DESCRIPTION:** SVL 70 mm. Head and body weakly depressed, with long limbs and a very long slender tail. Body scales mostly homogeneous; small uniform dorsal scales, each with a keel converging back towards midline. Scales on flanks smaller and smooth to weakly keeled. There are no crests, spines or enlarged scales. Pattern is dominated by a pair of complete to broken pale dorsolateral stripes. A total of 42–68 femoral and preanal pores extend along full length of thighs and curve sharply forwards at midline.

**SUBSPECIES:** *C. i. isolepis* is bright reddish brown above with dark dots, dark-edged pale spots, a pair of dark-edged yellow dorsolateral stripes, an upper lateral series of large dark brown spots (coalesced to form a stripe on males) and a pale midlateral stripe. Mature males have extensive dark ventral pigment: a broad black stripe from chin onto abdomen extending onto arms.

*C. i. gularis* (Sternfeld, 1924), has dorsolateral stripe breaking into spots or elongate blotches at anterior or mid body. Adult males have extensive black on the flanks and the black ventral pigment extends to the edges of the lower jaw.

*C. i. citrinus* (Storr, 1965) is dull to bright yellow above and the dorsolateral stripe breaks at anterior or mid body. Adult males have extensive black on the flanks and there is a broad black stripe from chin onto abdomen extending onto arms.

**KEY CHARACTERS:** Differs from the Spotted Sand Dragon (*C. maculatus*), the Long-tailed Sand Dragon (*C. femoralis*) and the Mallee Sand Dragon (*C. fordi*) in having preanal pores curving sharply forward at midline (versus slightly forwards). Differs from the Rufus Military Dragon (*C. rubens*) and further from *C. femoralis* in having a dark upper lateral zone.

**DISTRIBUTION AND ECOLOGY:** Arid to semi-arid shrublands and grasslands on sandy to loamy substrates, including spinifex deserts. In the Simpson Desert it is restricted primarily to sites providing more than 30% cover of hard spinifex (*Triodia basedowii*) (Daly *et al*, 2008). *C. i. isolepis* occurs from north-western WA to northern NT. *C. i. citrinus* occurs on yellow sandplains in the interior of southern WA. *C. i. gularis* extends across the vast deserts from the interior of WA to south-western Qld.

**BIOLOGY:** Extremely swift and wholly terrestrial, foraging in open spaces and around the edges of low vegetation, and dashing with mind numbing speed on all four limbs ahead of any potential approaching danger. It does not utilise any elevated perches, and shelters under vegetation rather than in burrows. Thermoregulatory behaviour includes using

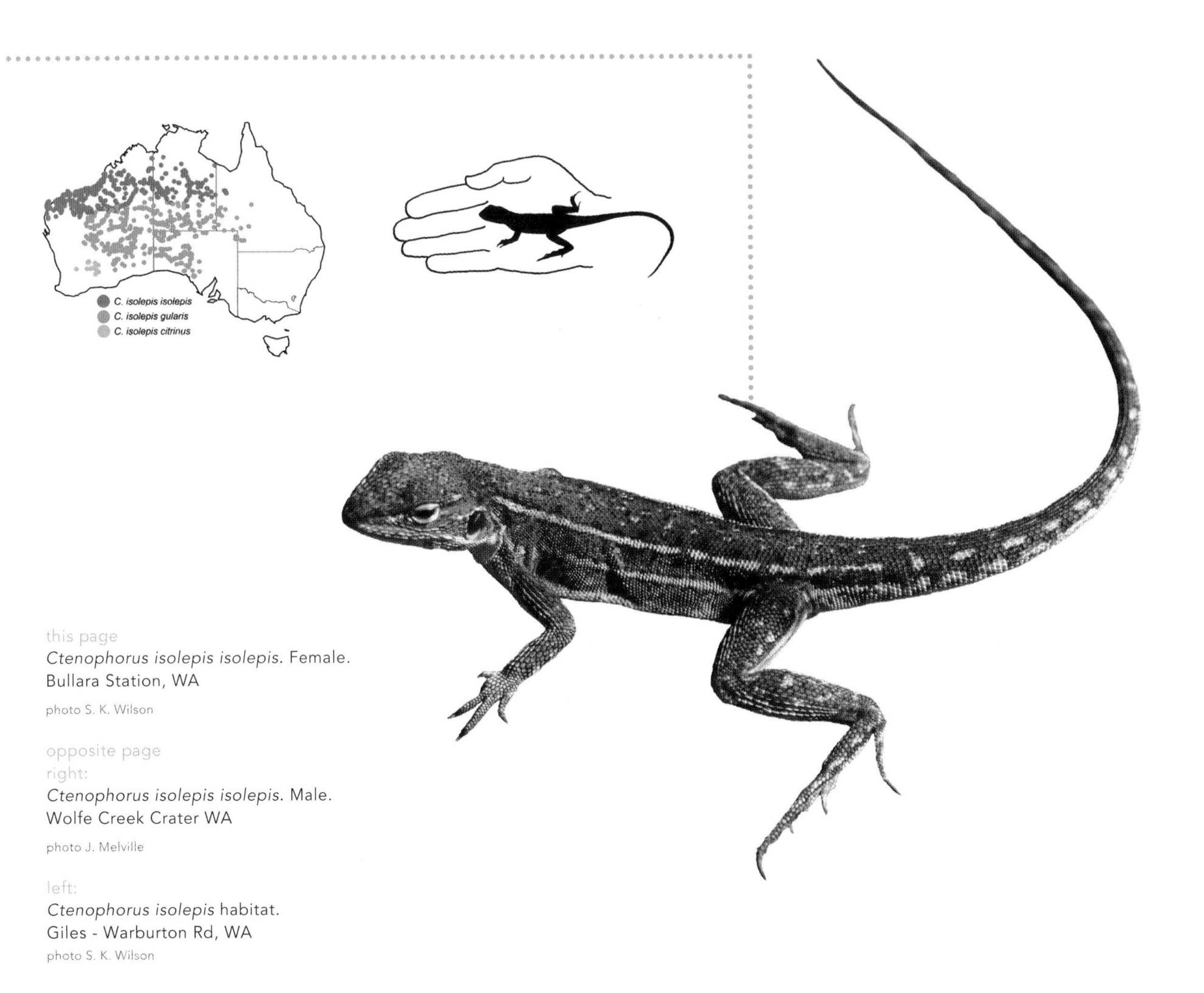


this page
*Ctenophorus isolepis isolepis*. Female.
Bullara Station, WA
photo S. K. Wilson

opposite page
right:
*Ctenophorus isolepis isolepis*. Male.
Wolfe Creek Crater WA
photo J. Melville

left:
*Ctenophorus isolepis* habitat.
Giles - Warburton Rd, WA
photo S. K. Wilson

the shade of spinifex (Melville & Schulte, 2001) and strategic posturing. On cool mornings the body is flattened to the ground, and as sand temperatures increase the body, tail and hind toes are held off the ground with front legs extended (Losos, 1987). Diet consists almost exclusively of small ants less than 5mm in length (Daly *et al*, 2008), though some vegetation is recorded, probably ingested incidentally. Clutches of 1–6 eggs are recorded, and 2 – 3 clutches may be laid between September and February. Hatchlings grow rapidly and mature at 6–9 months of age. (Pianka, 1971a). Lifespan is probably annual.

top:
*Ctenophorus isolepis gularis.* Male. Windorah, Qld
photo S. K. Wilson

bottom:
*Ctenophorus isolepis gularis.* Female. Hay River, NT
photo S. K. Wilson

opposite page:
*Ctenophorus isolepis citrinus.* Female. Ghooli, WA
photo S. K. Wilson

## SPOTTED SAND DRAGON

*Ctenophorus maculatus* Gray, 1831

**DESCRIPTION:** SVL 67 mm. Head and body weakly depressed, with long limbs and a very long slender tail. Body scales mostly homogeneous; small uniform dorsal scales, each with a keel converging back towards midline. Scales on flanks smaller and smooth to weakly keeled. There are no crests, spines or enlarged scales. Pattern is dominated by a pair of complete to broken pale dorsolateral stripes. A total of 38–58 femoral and preanal pores extend along full length of thighs and curve slightly forwards at midline.

**SUBSPECIES:** *C. m. maculatus* is brown with continuous to broken cream dorsolateral and pale grey midlateral stripes. Males have a small black patch on chin, spots on lower lips, a chevron on the throat and kite-shape on the chest

*C. m. griseus*, is grey with the dorsolateral stripes usually flushed anteriorly with red. Male ventral pattern is similar to that of *C. m. maculatus*, but with chevron more angular, its arms prolonged further forward like a tuning fork, and the patch on the chest is anchor-shaped. Largest race.

*C. m. badius* is reddish with yellow dorsolateral stripes. Males have a dark chevron on the throat and a bar, broken or constricted at midline, across the chest. Smallest race,

*C. m. dualis* males resemble *C. m. griseus*. Females are drab reddish brown with more obscure pattern.

**KEY CHARACTERS:** Differs from the Central Military Dragon (*C. isolepis*) and Rufus Military Dragon (*C. rubens*) in having preanal pores curving slightly forwards at midline (versus curving sharply forward). Differs from Long-tailed Sand Dragon (*C. femoralis*) and Mallee Sand Dragon (*C. fordi*) in having femoral pores extending full length of thigh (versus extending only to middle or three quarters of thigh). Differs further from *C. fordi* in having a black chevron (versus spots) on throat.

**DISTRIBUTION AND ECOLOGY:** Arid to semi-arid grasslands and shrublands on sandy substrates in WA. *C. m. maculatus* occupies white to pinkish sands along lower west coast. *C. m. badius* occurs on red sandy loams and dunes and pale coastal sands further north on mid-west coast; *C. m. griseus* occurs in heathlands and open woodlands across southern interior; and *C. m. dualis* occupies a narrow band of mallee and spinifex on the southern edge of the Nullarbor Plain.

**BIOLOGY:** Extremely swift and wholly terrestrial, foraging in open spaces and around the edges of low vegetation, and dashing swiftly on all four limbs ahead of any potential approaching danger. It does not utilise any elevated perches, and shelters under vegetation rather than in burrows. Diet probably comprises almost exclusively ants and lifespan is probably annual.

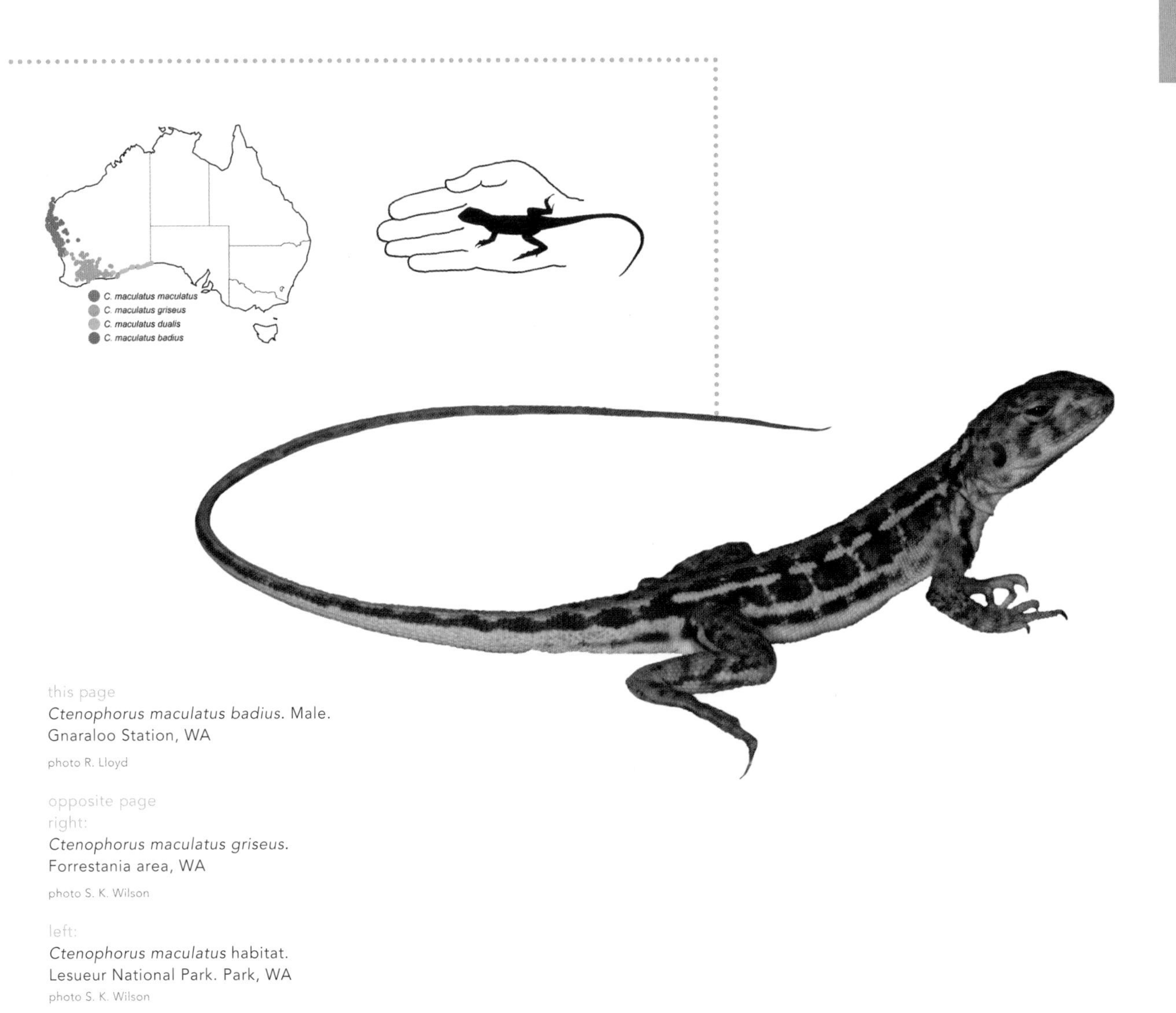


this page
*Ctenophorus maculatus badius*. Male.
Gnaraloo Station, WA

photo R. Lloyd

opposite page
right:
*Ctenophorus maculatus griseus*.
Forrestania area, WA

photo S. K. Wilson

left:
*Ctenophorus maculatus* habitat.
Lesueur National Park. Park, WA
photo S. K. Wilson

**NOTES:** *C. maculatus* belongs to the sand dragon species complex, a group closely allied to the military dragons (*C. isolepis* and *C. rubens*). These sand dragons, which also include *Ctenophorus fordi* and *Ctenophorus femoralis*, consist of 11 lineages occupying a range of sandy habitats (Edwards *et al*, 2015). Taxonomic work is underway.

top:
*Ctenophorus maculatus maculatus.* Tamala Station area, WA
photo S. K. Wilson

bottom:
*Ctenophorus maculatus badius.* Female. Gnaraloo Station, WA
photo R. Lloyd

opposite page:
*Ctenophorus maculatus dualis.* Border Village area, SA
photo S. Scott

## LAKE EYRE DRAGON

*Ctenophorus maculosus* Mitchell, 1948

**DESCRIPTION:** SVL 69 mm. Robust with short, deep head and relatively short limbs and tail. Scales mostly homogeneous: small and smooth with no nuchal or vertebral crests, and a few slightly enlarged scales tending to align transversely. Ears completely covered by scales, their presence indicated by depressions. White to very pale brown with fine dark reticulum or flecking, two rows of large, prominent circular black blotches between nape and hips, and a black streak running from chin to gular fold. The eye is deeply indented and rimmed with black. Red ventral flushes are acquired by both sexes to varying degrees during the breeding season.

**KEY CHARACTERS:** Differs from all sympatric *Ctenophorus* in having the ears completely covered by scales. Differs from *Typmanocryptis* species in having smooth, homogeneous body scales (versus numerous scattered to longitudinally aligned enlarged tubercles) and possessing a row of enlarged scales curving below eye.

**DISTRIBUTION AND ECOLOGY:** Restricted to featureless salt lakes in arid interior of SA, inhabiting the crusty salt along the shores of Lakes Eyre, Torrens, Callabonna and Frome. It is a unique and harsh environment, devoid of shade and fresh water, regularly buffeted by wind and dust storms, and baking under summer temperatures often exceeding 40°C. The only available elevated perching sites are raised salt crusts, the rims of ant nests, sparse driftwood and occasional animal bones.

**BIOLOGY:** Lake Eyre Dragons exhibit extraordinary modifications to survive in such an extreme environment. Their pale colours help reflect excess heat while fringed black eyelids reduce glare and help protect from grit during dust storms. And during periodic flooding the entire lizard population must migrate to the sandy shorelines.

The dragons excavate burrows under the salt crust, where a fine layer of silt offers a humid, stable protection from the heat. They feed on harvest ants (*Melophorus* spp.) and occasional insects blown by winds onto the lake surface.

Dominant males occupy the choicest perching sites and engage in aggressive territorial contests with each other (McLean & Stuart-Fox, 2015). They are also active during the most favourable times, while subservient males make do with inferior vantage points, and their activity options are largely restricted to less favourable, hotter times of day (Mitchell, 1973).

Males persistently and aggressively attempt to mate, and females are even occasionally killed (Olsson, 1995). To avoid unwanted advances, large females may threaten with an arched back and laterally compressed body. Smaller ones simply

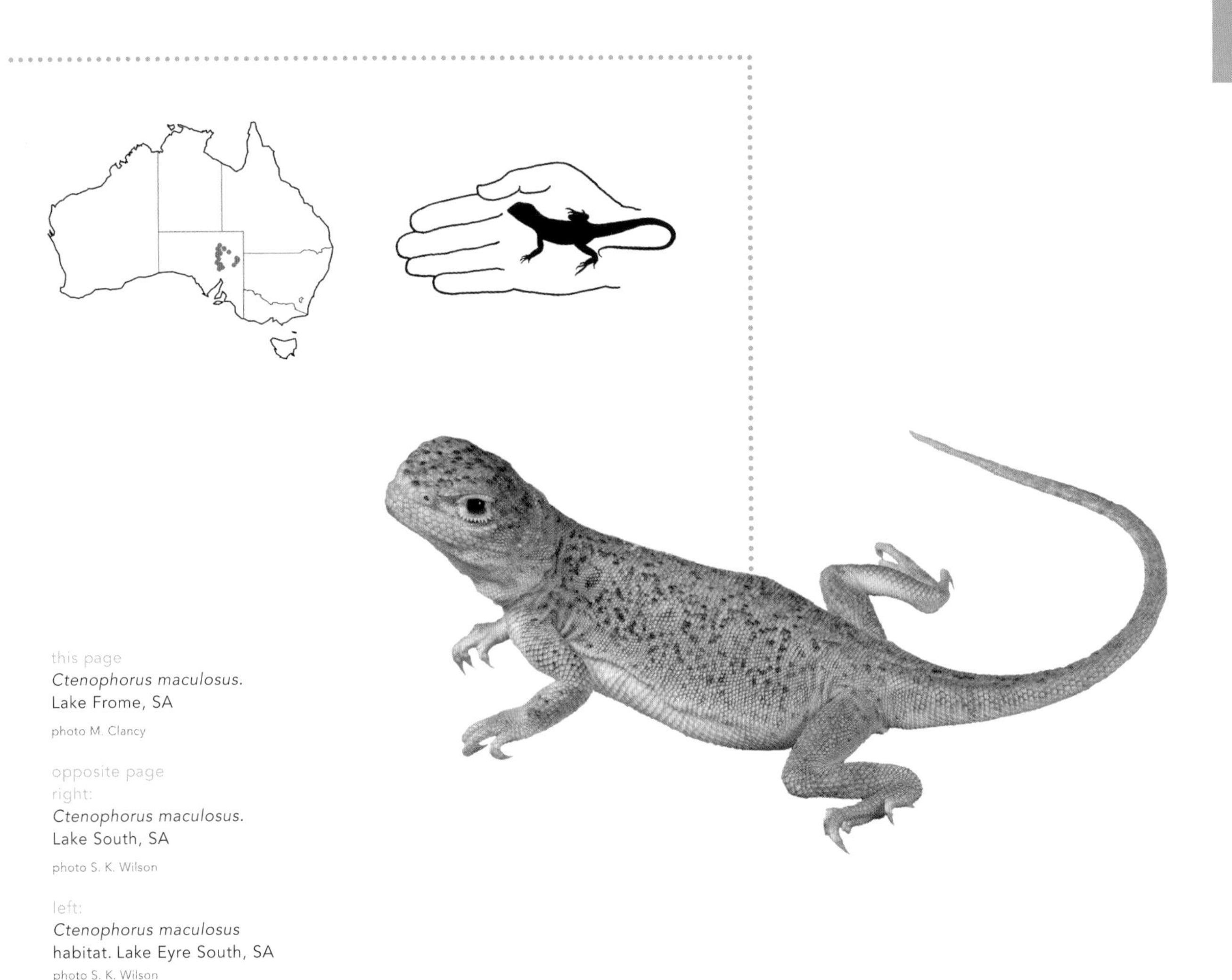

this page
*Ctenophorus maculosus.*
Lake Frome, SA
photo M. Clancy

opposite page
right:
*Ctenophorus maculosus.*
Lake South, SA
photo S. K. Wilson

left:
*Ctenophorus maculosus*
habitat. Lake Eyre South, SA
photo S. K. Wilson

flee. Females also adopt a unique posture to discourage mating, rolling onto their backs to lie belly up, displaying a bright red abdomen. The males cannot mate with them in this position, so they may pin them to the ground to prevent them flipping over (McLean *et al*, 2010). The intensity of the bright red colour of females and their behaviour towards males has been linked to sex hormone levels, with the brightest colouration at ovulation and the strongest rejection of males when they are gravid (Jessop *et al*, 2009).

*Ctenophorus maculosus.*
Lake South, SA
photo S. K. Wilson

## MCKENZIE'S DRAGON

*Ctenophorus mckenziei* Storr, 1981

**DESCRIPTION:** SVL 77 mm. Head moderately long and narrow, body is slender and limbs and tail relatively long. Body scales small and uniform with nuchal crest of compressed triangular scales continuous with a keeled vertebral line to base of tail, and dorsolateral spines on neck. Blackish brown with narrow pale vertebral stripe, broad wavy-edged and sometimes broken orange brown dorsolateral stripes and irregular narrow pale transverse bands. On breeding males there is a broad black stripe on throat and a black patch on chest extending to forearms and abdomen and sometimes a dark patch anterior to vent. A total of 36–48 preanal and femoral pores curving forward but well separated at midline.

**KEY CHARACTERS:** Differs from the Crested Dragon (*C. cristatus*) in lacking dorsolateral crests and from Lozenge-marked Dragon (*C. scutulatus*) in having consistently darker dorsal colouration and attaining smaller size (SVL 77 mm versus 115 mm).

**DISTRIBUTION AND ECOLOGY:** Semiarid southern woodlands of mallee and Acacia growing over chenopod shrubs at two widely-spaced localities: near Noondoonia Station and Ponier Rock in south-eastern interior of WA, and Colona Station area, south-western SA.

**BIOLOGY:** Ants and termites are significant components of a broad arthropod diet. Rival males have been recorded to head-bob, wave forearms, and present to each other by facing the same direction, loosely coiling their tails and pushing up with the hind limbs (Peterson *et al*, 1994).

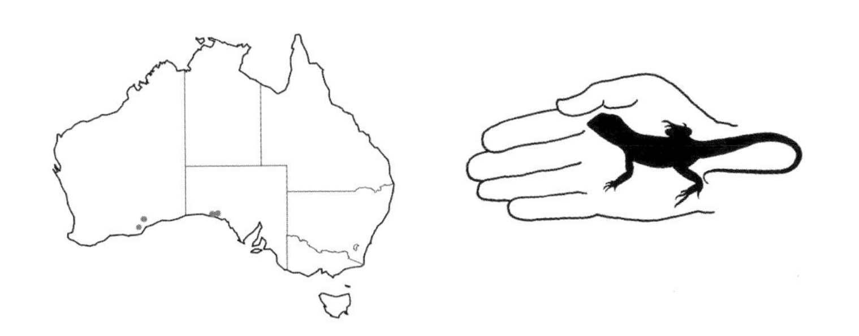

opposite page:
*Ctenophorus mckenziei* habitat.
Noondoonia Station, WA
photo B. Maryan

below:
*Ctenophorus mckenziei.*
Noondoonia Station, WA
photo R.Browne-Cooper

## BARRIER RANGE DRAGON

*Ctenophorus mirrityana* McLean, Moussalli, Sass and Stuart-Fox, 2013

**DESCRIPTION:** SVL 91mm. Head and body dorsally depressed. Scales on snout smooth to lightly wrinkled, and scalation on body mostly homogeneous; small uniform scales, each with a low keel, and a low crest of enlarged, strongly keeled scales on nape. Colouration differs markedly between the sexes. Adult males blue grey to pale blue with a broad black upper lateral stripe from behind eye to groin. Enclosed within this is a thinner, broken orange stripe. There is an orange wash over eyes, snout and upper jaw, and the throat is cream with an orange flush, parallel grey stripes and a broad black midline stripe from snout to gular fold. Females and juveniles are brown with a thin pale vertebral line, dark lateral stripe and brown, grey and terracotta speckling. Ventrally they are grey to cream with grey stripes on the throat and, during the breeding season, an orange flush on the belly. Femoral and preanal pores 34–42.

**KEY CHARACTERS:** Differs from the Tawny Dragon (*C. decresii*) mainly by male colouration: a black median stripe on throat present (versus absent), and a broken orange line present within the black lateral stripe (versus variable, but black lateral stripe edged above with yellow to orange or red).

**DISTRIBUTION AND ECOLOGY:** Exclusively rock inhabiting, on habitats ranging from outcrops, gorges and ranges to stony road spoils in western NSW from the Silverton and Broken Hill areas to Mutawintji National Park, and Koonenberry Mountain. The flat head and body allow the dragons to shelter in narrow rock crevices.

**BIOLOGY:** Courtship and territorial displays of males involve push ups and tail-flicks, generally from a prominent, elevated site. Clutches of 4–7 eggs are recorded. The sex of offspring is determined by temperature, with low incubation temperatures of 25°C producing all females (Harlow & Price, 2000, as *C. decresii*).

**NOTES:** In a review of reptiles of particular conservation concern in western NSW, *C. mirrityana* (then *C. decresii*) was identified as a species at risk because of its restricted distribution (Sadlier & Pressy, 1994) and is currently listed as Near Threatened by the IUCN (Melville *et al*, 2017b). The rocky terrain where the species occurs is grazed by feral goats. Grazing pressure from goats, which can be extreme during droughts, is believed to be of significant threat to *C. mirrityana*.

this page
*Ctenophorus mirrityana*. Male.
Mutawintji National Park, NSW

photo S. K. Wilson

opposite page
right:
*Ctenophorus mirrityana*. Male.
Mutawintji National Park, NSW

photo S. K. Wilson

left:
*Ctenophorus mirrityana*
habitat Mutawintji National
Park, NSW
photo G Swan

*Ctenophorus mirrityana*. Female.
Mutawintji National Park, NSW
photo S. K. Wilson

## LAKE DISAPPOINTMENT DRAGON

*Ctenophorus nguyarna* Doughty, Maryan, Melville and Austin, 2007

**DESCRIPTION:** SVL 78 mm. Moderately robust with short, deep head and moderately short limbs and tail. Body scales heterogeneous; small and conical on flanks with numerous enlarged scales that are scattered towards midline and tend to form transverse bars on sides. There are a few slightly enlarged tubercles along skin folds on sides of neck and an erectable nuchal crest continuous with a vertebral crest of weakly keeled scales. Sexes differ. Adult males are orange with black variegations and numerous small dark-edged pale spots. A broad silvery grey vertebral stripe extends from nape onto tail, enclosing sharp-edged black bars, often off-set at midline. The sides of the tail are marked with black vertical bars. Ventral surfaces have a black triangle on throat and kite-shaped patch on chest. Females are duller greyish orange with weaker, narrower vertebral stripe and no dark ventral markings. Femoral pores 16–18. Preanal pores 9–12.

**KEY CHARACTERS:** Differs from the Claypan Dragon (*C. salinarum*) in having enlarged paler scales tending to form transverse bars on flanks (versus enlarged scales tend to form transverse bars on back near midline).

**DISTRIBUTION AND ECOLOGY:** Known only from Lake Disappointment, a large saline basin in the Little Sandy Desert, WA. Burrows are excavated at the bases of samphire shrubs (*Halosarcia halocnemoides*). It appears to be closely tied to shrubs fringing the lake (Doughty *et al*, 2007).

**BIOLOGY:** Burrows are like those of its close relatives, the Painted Dragon (*C. pictus*) and Claypan Dragon (*C. salinarum*); generally U-shaped with an escape exit terminating near the surface. Males are often observed on the crowns of samphire clumps. From these vantage points they engage in displays featuring head bobbing, vigorous back-arching and tail-waving. Regular spacing of males suggests they maintain exclusive territories. The more terrestrial females do not perch on the shrubs. The dragons have been observed to feed on harvester ants. Clutches of 1–4 eggs are recorded, probably laid in spring.

**NOTES:** Listed as a conservation priority species in WA and as vulnerable on the IUCN Redlist (Catt *et al*, 2017). A proposed potash mine and associated road-building near the only known locality is a potential threat to this species.

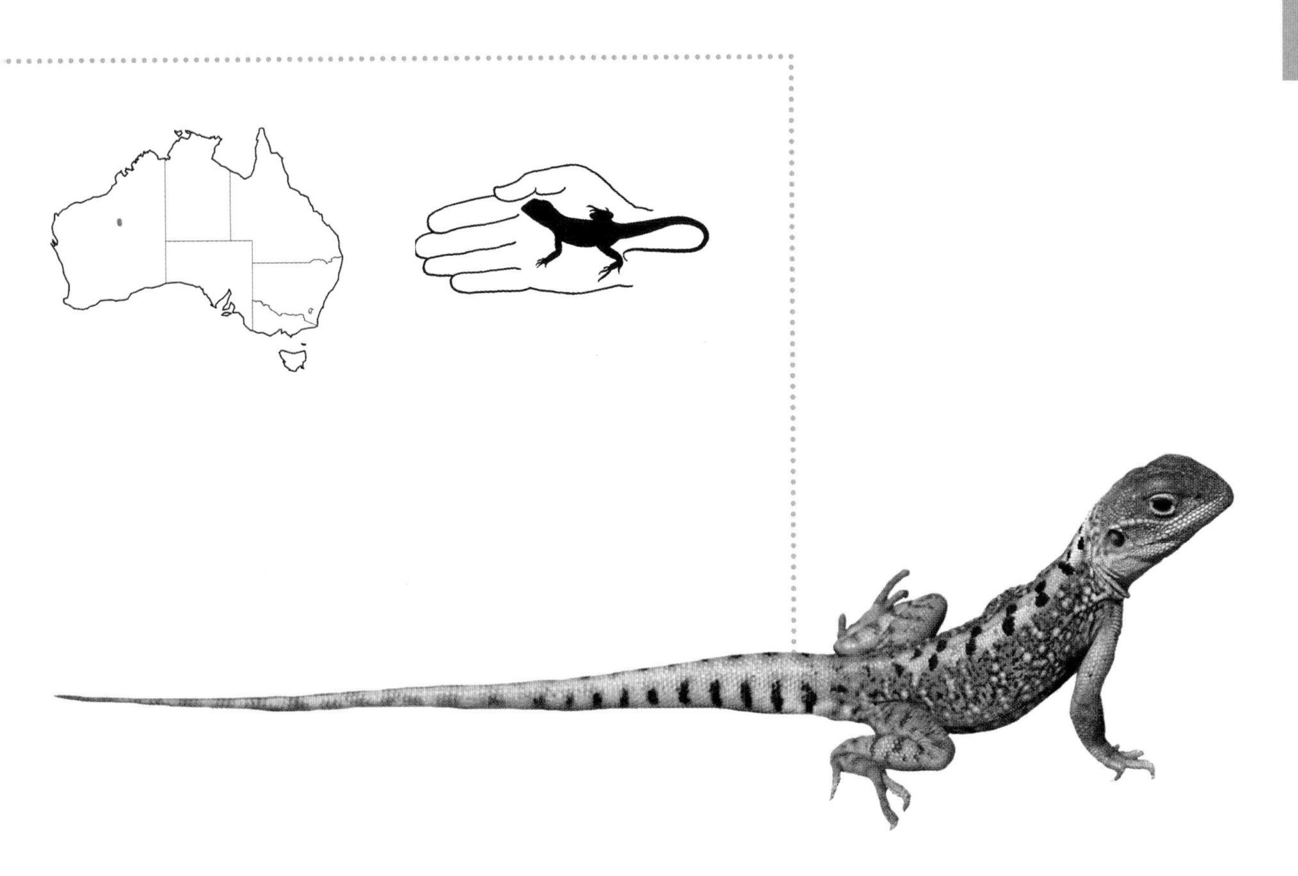

this page
*Ctenophorus nguyarna.*
Lake Dispappointment, WA

photo B. Maryan

opposite page
right:
*Ctenophorus nguyarna.*
Lake Disappointment, WA

photo G. Gaikhorst

left:
*Ctenophorus nguyarna* habitat.
Lake Dispappointment, WA
photo B. Maryan

## CENTRAL NETTED DRAGON

*Ctenophorus nuchalis* De Vis, 1884

**DESCRIPTION:** SVL 115 mm. Moderately large and robust with round head, very blunt snout, short limbs and tail. Scales mainly homogeneous; mostly flat, sometimes with an enlarged raised scale enclosed within pale markings over back and sides, some sharp conical tubercles on sides of neck and a small spiny nuchal crest. Pale yellowish brown with dark reticulum and narrow pale vertebral stripe. Breeding males are flushed with bright orange-red over the head and throat. Femoral and preanal pores 12–34 arranged in a curve, sweeping forward to anterior thigh.

**KEY CHARACTERS:** Differs from the Western Netted Dragon (*C. reticulatus*) and Gibber Dragon (*C. gibba*) in having femoral pores curving forward (versus in a straight line along thigh), and further from the Gibber Dragon in lacking a black line and blotch on throat.

**DISTRIBUTION AND ECOLOGY:** Extremely widespread across Australia's semiarid to arid areas, in diverse habitats ranging from sand plains and dunes vegetated with spinifex to *Acacia* woodlands and shrublands on heavy loams and stony soils. It is considered to be relatively tolerant to impacts from grazing (Frank *et al*, 2013), prefers habitats with sparse (less than 10%) vegetation cover (Daly *et al*, 2008), and has been found to occur in high numbers immediately following fires (Letnic *et al*, 2004) . Each individual excavates several shallow, sloping burrows up to 50 centimetres long at bases of shrubs and stumps, and in windrows beside vehicle tracks. Burrow location is dependant on the availability of both perching sites and shade. During winter inactivity these are plugged with soil.

**BIOLOGY:** This dragon is abundant, and is often one of the most conspicuous lizards throughout the interior. They bask on low elevated sites such as rocks, stumps, termitaria and discarded roadside tyres, dashing on all four limbs to the nearest burrow when approached. *C. nuchalis* spends most of its active time in open sun, with body temperatures of 36°–48°C (Daly *et al*, 2007). Studies at Shark Bay in WA reveal most individuals are annual, with less than two per cent of more than 1000 marked animals surviving longer than one year (Bradshaw, 1986). The breeding cycle is flexible and dictated by rainfall, with breeding after winter rains in the southern part of its range, and in late summer following cyclonic rains in the Pilbara region, where it may also breed in spring if rains are substantial (Dickman *et al*, 1999). Although their diet is primarily arthropods, they consume a higher proportion of plant matter than smaller species in the genus (Greer, 1989). Clutches of 2–6 eggs are recorded (Greer, 1989). As an abundant and prolific burrower, this species is an important contributor of shelter sites for a range of other organisms. A host of animals from geckos and small snakes to invertebrates regularly utilise Central Netted Dragon burrows.

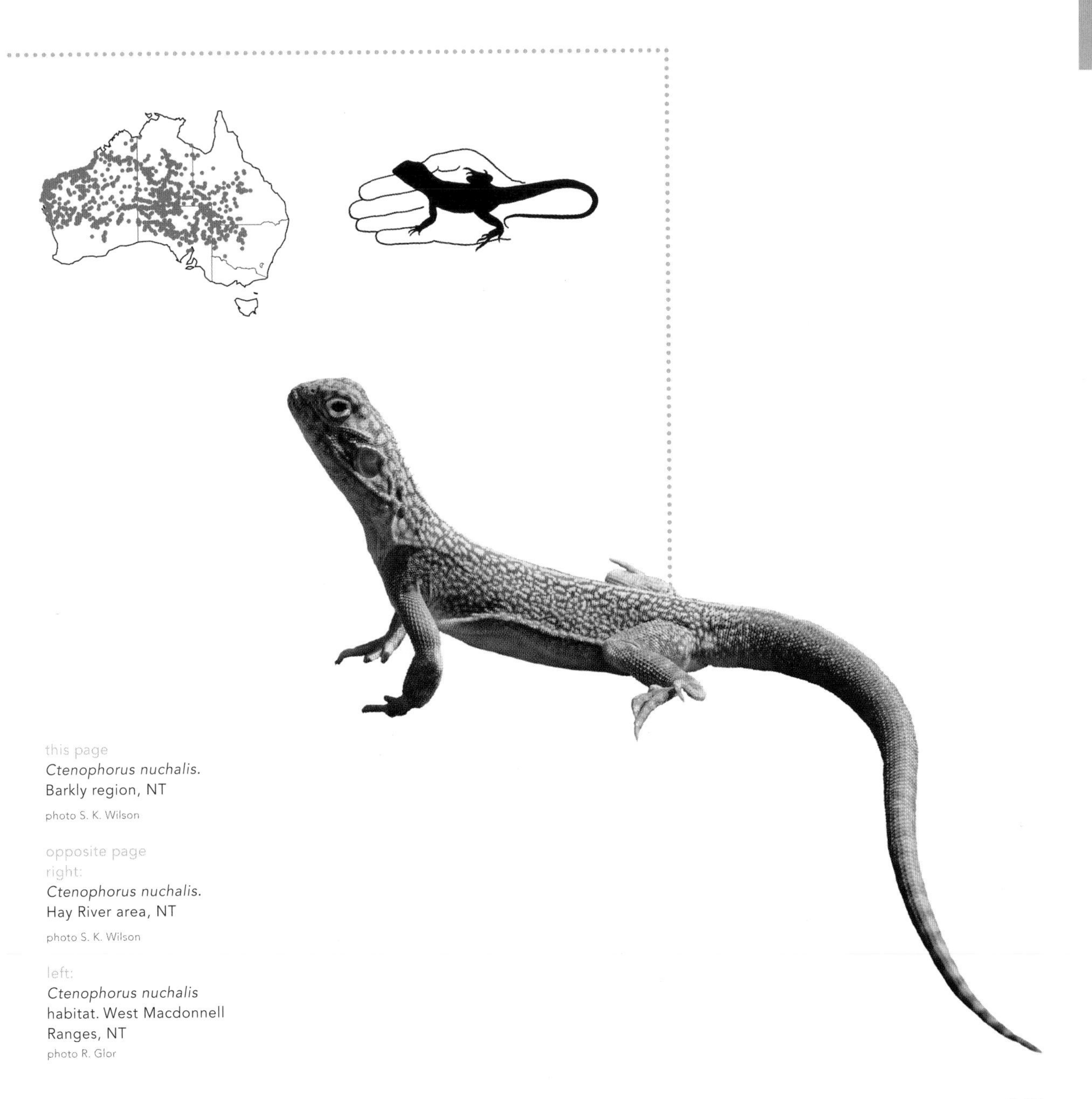

this page
*Ctenophorus nuchalis.*
Barkly region, NT
photo S. K. Wilson

opposite page
right:
*Ctenophorus nuchalis.*
Hay River area, NT
photo S. K. Wilson

left:
*Ctenophorus nuchalis*
habitat. West Macdonnell
Ranges, NT
photo R. Glor

*Ctenophorus nuchalis.*
Quilpie area, Qld.
photo S. K. Wilson

## ORNATE DRAGON

*Ctenophorus ornatus* Gray, 1845

DESCRIPTION: SVL 93 mm. Head and body extremely dorsally depressed. Scalation on body mostly homogeneous; small uniform scales, each with a low keel converging towards midline, and sometimes a low crest of enlarged scales on nape. Enlarged spiny scales present below, behind and in front of ear. Colouration varies between sexes, particularly among adult males, according to climate and substrate. Males on Darling Range and south coast are black to blackish brown with sharply contrasting vertebral series of pale blotches and spots. Further inland into semi-arid zones the pale blotches contrast with a reddish brown ground colour. In the more arid zones, from Paynes Find northwards, blotches are replaced by a prominent, dark-edged pale vertebral stripe against a pinkish red background. Males of all populations have prominent dark and pale rings around tail. Females and juveniles are brown to grey, typically with large pale blotches or ocelli along vertebral region, transverse rows of small pale dots, and obscure bands on tail. Femoral and preanal pores 22–34.

KEY CHARACTERS: The extremely dorsally depressed head and body distinguish Ornate Dragons from all sympatric species.

DISTRIBUTION AND ECOLOGY: Endemic to south-western WA, from the Darling Range near Perth, east to about Cape Le Grand and some islands in the Archipelago of the Recherche on the south coast, and north through the Goldfields and Murchison regions. Restricted to granite, occupying bare expanses strewn with exfoliations and boulders. The extremely flat build allows access to narrow rock crevices. Lizards bask on elevated rocks and sprint rapidly, on all four limbs across open rock sheets.

BIOLOGY: Social and thermoregulatory behaviour has been well studied. Lizards emerge from shelter with a body temperature of around 24°C, bask with backs oriented to the sun and bodies flattened against the rock, and become active once their body temperature is 37°C (Bradshaw & Main, 1968). While dorsal colours generally match the various granite backdrops, throat reflectance in the ultraviolet spectrum contrasts strongly. This provides an important signalling mechanism via head-bobbing and push-up displays, to communicate sexual status and receptivity (Le Bas & Marshall, 2000). Clutches of 2–5 eggs are recorded (Greer, 1989) and the sex of offspring is probably determined by temperature, with low incubation temperatures of 25°C producing all females (Harlow & Price, 2000).

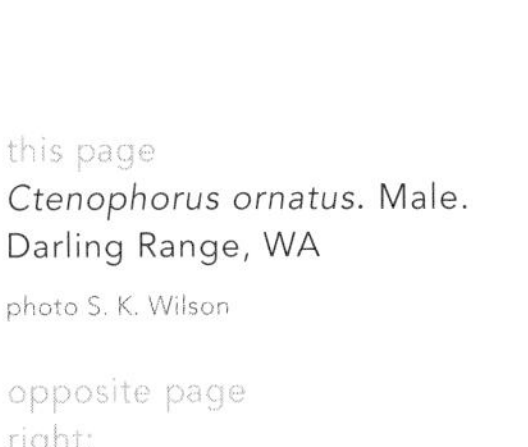

this page
*Ctenophorus ornatus.* Male.
Darling Range, WA
photo S. K. Wilson

opposite page
right:
*Ctenophorus ornatus.* Male.
Charles Darwin Nature Reserve. WA
photo S. K. Wilson

left:
*Ctenophorus ornatus* habitat.
Charles Darwin Nature Reserve. WA
photo S. K. Wilson

**NOTES:** Genetic research has found that there are 23 genetic subgroups within two major evolutionary linages of *C. ornatus*, reflecting both ancient and contemporary isolation of populations on granite outcrops (Levy *et al*, 2012). Researchers emphasise the importance of these two evolutionarily lineages in the conservation management of these lizards. Land clearing restricts movement between rocky outcrops, thus reducing genetic variation with potentially negative consequences for the long-term persistence of populations (Levy *et al*, 2010). Genetic evidence has also found that this species hybridizes with *C. caudicinctus*, both in the north of its range and around Higginsville in the Goldfields (Melville *et al*, 2017).

*Ctenophorus ornatus*. Male.
Charles Darwin Nature Reserve, WA
photo S. K. Wilson

*Ctenophorus ornatus*. Female.
Ravensthorpe area, WA
photo S. K. Wilson

## NORTH-WESTERN HEATH DRAGON

*Ctenophorus parviceps* Storr, 1964

**DESCRIPTION:** SVL 46 mm. Very small with short limbs and tail. Body scales heterogeneous; uniform and weakly spinose down middle of back, but varying greatly in size on dorsolateral area, including erect tubercles mainly restricted to areas of dark colour. Tympanum completely covered by scales. Chin 'terraced' (a straight line forming angular junction along ventrolateral edge of jaw). Weakly patterned. Pale brown along vertebral region and broad darker brown dorsal zone enclosing pale, longitudinally elongate hourglass-shaped blotches. Femoral and preanal pores 26–34.

**KEY CHARACTERS:** Differs from the Shark Bay Heath Dragon (*C. butlerorum*) by its weaker pattern, absence of yellow on chin and more femoral and preanal pores (26–34 versus 14–20).

**DISTRIBUTION AND ECOLOGY:** Restricted to a strip along the mid-west coast of WA, between North West Cape and Carnarvon, and off-shore on Bernier Island. Occupies coastal dunes composed of sand or shell grit, and vegetated with beach spinifex (*Spinifex longifolius*) and low shrubs. Not particularly swift when compared to other small dragons, tending to rely mainly on its cryptic disruptive pattern to avoid predators. When pursued it scuttles rather than sprints to the cover of low vegetation.

**BIOLOGY:** Ants probably form a major component of the diet.

opposite page:
*Ctenophorus parviceps* habitat.
North West Cape, WA
photo S. K. Wilson

below:
*Ctenophorus parviceps*.
Yardie Creek, WA
photo S. K. Wilson

## PAINTED DRAGON

*Ctenophorus pictus* Peters, 1866

**DESCRIPTION:** SVL 65mm. Moderately robust with short, deep head and moderately short limbs and tail. Body scales homogeneous; flat and weakly keeled dorsally, grading into smaller scales on flanks. There are a few slightly enlarged tubercles along skin folds on sides of neck and a weak erectable nuchal crest continuous with a vertebral crest of enlarged keeled scales. Extremely variable, and sexes differ. Adult males are brown, yellowish brown to orange with a broad dark bluish-grey vertebral stripe overlayed by dark-edged pale bars, blotches or spots. Breeding males develop bright head colours, including red, orange and yellow. Males also develop a blue flush over lower lips, throat and limbs, and bright yellow to orange over anterior chest and shoulders. Females and juveniles are duller, lacking blue and bright yellow pigment, although females may have some yellow pigmentation on their throats. Femoral and preanal pores 35–46 in a straight line along full length of thighs.

**KEY CHARACTERS:** Differs from the Claypan Dragon (*C. salinarum*) in having homogeneous body scales (versus mixed with enlarged paler scales).

**DISTRIBUTION AND ECOLOGY:** Semiarid to arid southern interior and south coast, on sandy soils vegetated with heaths or other shrubs, spinifex or cane grass. Burrows are excavated at the bases of vegetation, often in raised slopes of sand consolidated by the plants.

**BIOLOGY:** Burrows are generally U-shaped with an escape exit terminating near the surface. Painted Dragons are mainly terrestrial, resting only on very low perches. Social behaviour and the evolutionary significance of their bright colouration has been well studied. Males have been observed displaying to rivals by raising their nuchal crest and vertebral ridge, holding their bodies high off the ground and presenting their broadest lateral aspect. This is an impressive display by an extremely colourful dragon. Males with read heads have been found to be dominant and the most aggressive, while individuals with yellow heads actually sire more offspring (Olsson *et al*, 2012b). Males can also be divided into those that have colourful yellow throat "bibs" and those that don't, with bibbed males having higher levels of paternity (Olsson *et al*, 2009). Females are non-territorial and non-aggressive. Females potentially lay four, or possibly more, clutches of 2–6 eggs in spring and summer (Greer, 1989; Niejalke, 2006). *C. pictus* is an annual species in the wild, with about 80% of individuals dying within one year (Olsson *et al*, 2009).

opposite page left:
*Ctenophorus pictus* habitat.
Innamincka area, SA
photo S. K. Wilson

opposite page right:
*Ctenophorus pictus*. Breeding male.
Ballera area, Qld
photo S. K. Wilson

below:
*Ctenophorus pictus*. Breeding male.
Oodnadatta, SA
photo P. Robertson

*Ctenophorus pictus*. Female.
Ballera area, Qld.
photo S. K. Wilson

*Ctenophorus pictus*. Breeding male.
Moomba, SA.
photo S. K. Wilson

## WESTERN NETTED DRAGON

*Ctenophorus reticulatus* Gray, 1845

**DESCRIPTION:** SVL 108 mm. Moderately large and robust with round head, very blunt snout, short limbs and tail. Scales mainly homogeneous; mostly flat, with numerous enlarged flat scales enclosed within pale markings over back and sides, some sharp conical tubercles on sides of neck and a small spiny nuchal crest. Sexes differ. Adult males are reddish brown to red with black reticulum. Breeding males are deep red with reddish flush on chin and throat. Juveniles are olive-grey marked with dark paravertebral spots alternating with bands of small whitish spots. Females lose the pale spots and develop a series of elongate dark grey blotches. Femoral and preanal pores 30–56 arranged in a straight line along rear edge of thigh.

**KEY CHARACTERS:** Differs from the Central Netted Dragon (*C. nuchalis*) in having the femoral pores arranged in a straight line along rear of thigh (versus curving forward towards front of thigh). Differs from the Gibber Dragon (*C. gibba*) in lacking a black line and blotch on throat.

**DISTRIBUTION AND ECOLOGY:** Arid to semiarid west to central interior. They favour open Acacia-dominated woodlands and shrublands growing on heavy, often stony soils. Shallow burrows are excavated under rocks, logs, and at bases of shrubs.

**BIOLOGY:** These dragons bask on low, slightly elevated sites such as stones, stumps and woody shrubs, dashing on all four limbs to the nearest burrow when approached. They actively thermoregulate, with an average body temperature of 34°C (Pianka, 2014). An omnivorous species, consuming ants, termites, grasshoppers and a relatively higher proportion of plant material (about 24%) compared with smaller species in the genus. Clutches of 2–8 eggs have been recorded (Greer, 1989).

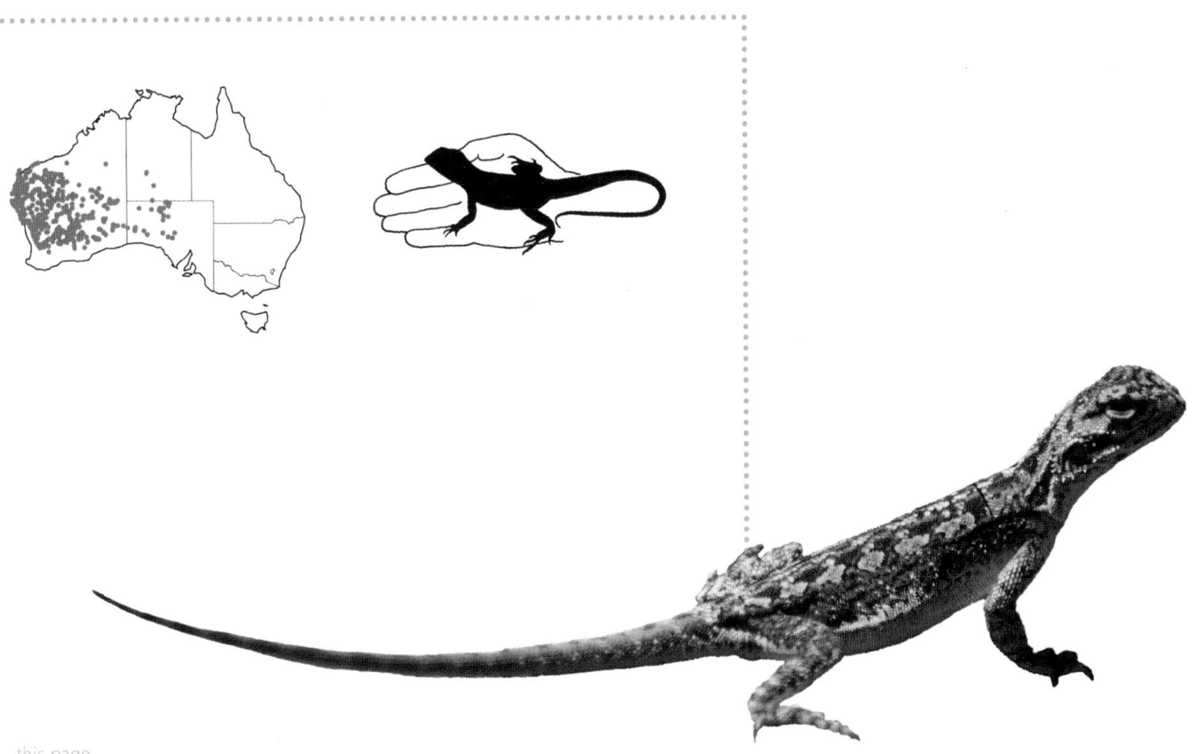

this page
*Ctenophorus reticulatus*. Female.
Ninghan Station WA
photo J. Melville

opposite page
right:
*Ctenophorus reticulatus*. Male.
Winduldara Rockhole, WA
photo S. K. Wilson

left:
*Ctenophorus reticulatus* habitat.
Meekatharra area, WA
photo S. K. Wilson

following page:
*Ctenophorus reticulatus*. Male.
Ninghan Station WA
photo J. Melville

## RUFUS MILITARY DRAGON

*Ctenophorus rubens* Storr 1965

**DESCRIPTION:** SVL 83 mm. Head and body weakly depressed, with long limbs and a very long slender tail. Body scales mostly homogeneous; small uniform dorsal scales, each with a keel converging back towards midline. Scales on flanks smaller and smooth to weakly keeled. There are no crests, spines or enlarged scales. Sexes differ. Mature males are pinkish brown flushed with brown on head and tail, with little or no pattern other than barely discernible pale dorsolateral stripes. Black ventral pigment is extensive over throat, chest, limbs, abdomen and sometimes vent. Females are dark reddish brown with pale dorsolateral and midlateral stripes, dark spots and dark-edged pale spots or bars. A total of 50–72 femoral and preanal pores extend along full length of thighs and curve sharply forwards at midline.

**KEY CHARACTERS:** Differs from the Spotted Sand Dragon (*C. maculatus*) and the Long-tailed Sand Dragon (*C. femoralis*) in having preanal pores curving sharply forward at midline (versus slightly forwards). Differs from the Central Military Dragon (*C. isolepis*) in lacking a dark upper lateral zone, and further from all in growing to a larger size (SVL 83 mm versus SVL 70 mm or less).

**DISTRIBUTION AND ECOLOGY:** Red sands with spinifex, also red loams and clays with spinifex, soft grasses and low shrubs. Distributed on the mid-west coast of WA, in vicinity of Exmouth Gulf, with an isolated population on sandplains south of Hamelin Pool.

**BIOLOGY:** Extremely swift and wholly terrestrial, foraging in open spaces and around the edges of low vegetation, mainly spinifex. For its size, this appears to be one of the fastest Australian lizards, sprinting on all four limbs ahead of any potential approaching danger. It does not utilise any elevated perches, and shelters under vegetation rather than in burrows. Its distribution overlaps with several similarly terrestrial dragon species, but it exhibits subtle differences in habitat choice. At Giralia Station it occurs adjacent to the Long-tailed Sand Dragon (*C. femoralis*). There, it occupies sand flats while *C. femoralis* lives on the dune crests and slopes (Wilson & Knowles, 1988). Diet probably comprises mostly ants and lifespan is probably annual.

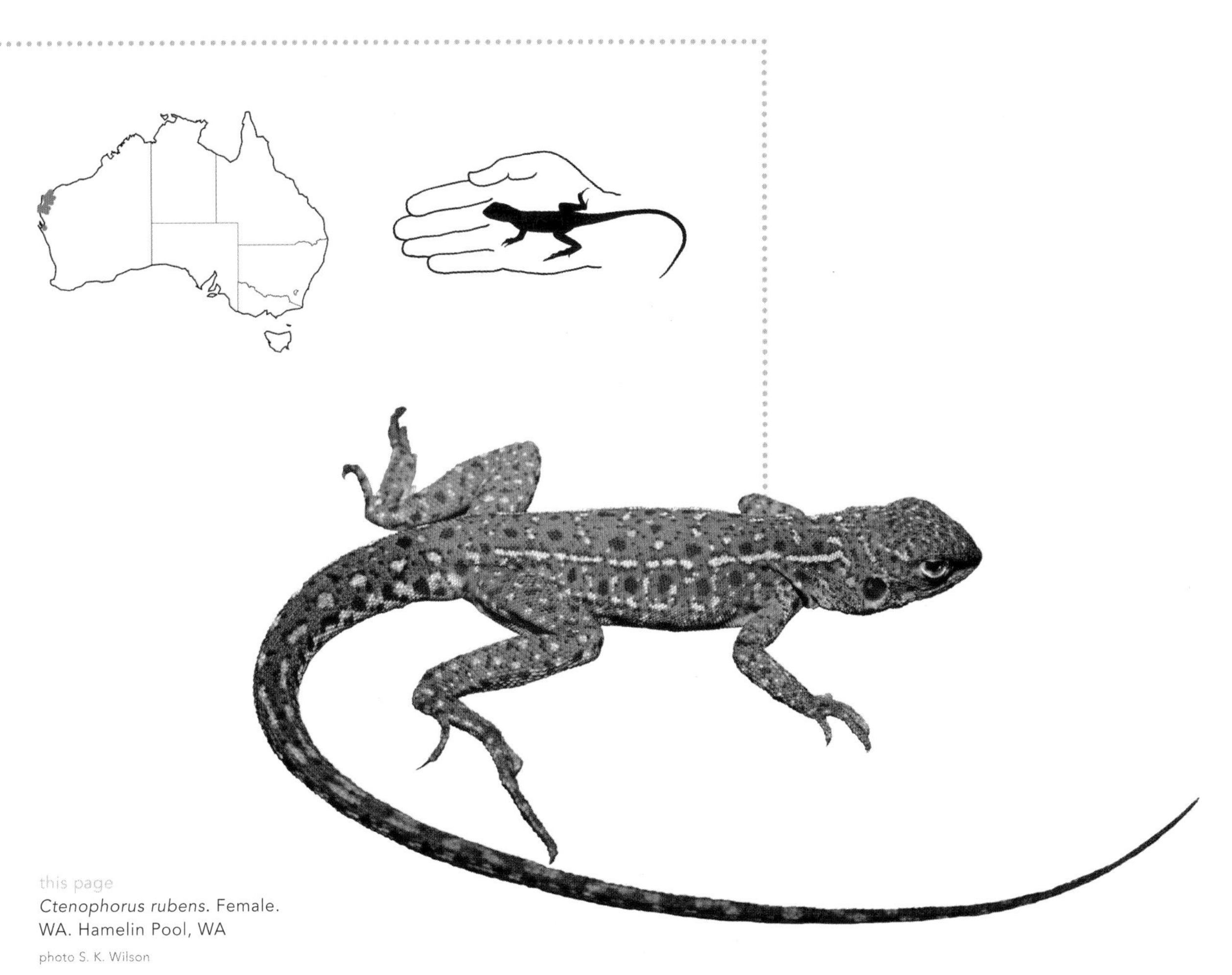

this page
*Ctenophorus rubens*. Female.
WA. Hamelin Pool, WA
photo S. K. Wilson

opposite page
right:
*Ctenophorus rubens*. Male.
Koodarie stn, WA
photo A. Elliott

left:
*Ctenophorus rubens* habitat.
Giralia Station, WA
photo S. K. Wilson

## RUSTY DRAGON

*Ctenophorus rufescens* Stirling and Zietz, 1893

**DESCRIPTION:** SVL 92 mm. Head and body extremely dorsally depressed. Scalation on body homogeneous; small weakly keeled scales with keels converging back towards midline, a low nuchal crest on males and no vertebral crest on body. Sexes differ. Mature males are pink to brown with brighter orange on flanks and tail but little or no pattern. Females and juveniles are rusty brown with series of black spots along either side of midline between nape and base of tail, an irregular series along upper flanks below dorsolateral fold and another along flanks. Femoral and preanal pores 40–63.

**KEY CHARACTERS:** The strongly flattened head and body distinguish *Ctenophorus rufescens* from all sympatric species.

**DISTRIBUTION AND ECOLOGY:** Exposed granite outcrops strewn with exfoliations and boulders in north-western SA, south-western NT and adjacent WA. The dragons perch on elevated sites, and if disturbed dash beneath a rock slab or into a crevice.

**BIOLOGY:** There is very little published data on the biology of this species. The extremely flattened head and body, approaching that of the Ornate Dragon (*Ctenophorus ornatus*), probably reflects a similar reliance on narrow gaps under exfoliations and boulders. The use of complex displays for social behaviour, a distinctive feature of several other SA rock dragons (Stuart-Fox *et al*, 2004), has not yet been recorded.

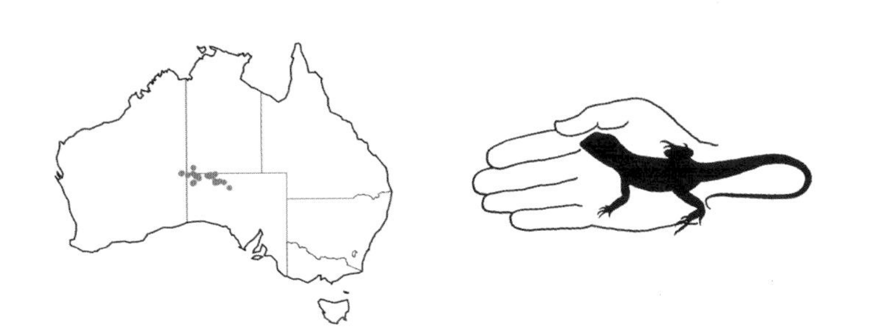

opposite page:
*Ctenophorus rufescens* habitat.
Musgrave Ranges, SA
photo J. de Jong

below:
*Ctenophorus rufescens.*
Musgrave Ranges, SA
photo J. de Jong

## CLAYPAN DRAGON

*Ctenophorus salinarum* Storr, 1966

DESCRIPTION: SVL 70 mm. Moderately robust with short, deep head and moderately short limbs and tail. Body scales heterogeneous; flat and weakly keeled, mixed with enlarged smooth paler scales tending to be arranged in transverse rows particularly towards midline. There are a few slightly enlarged tubercles along skin folds on sides of neck and a weak erectable nuchal crest continuous with a vertebral crest of enlarged keeled scales. Reddish brown to grey with dark-edged pale spots joined to form transverse bars along vertebral region, and tending to align with enlarged pale scales on flanks. Sexes similar, but breeding males are brightest, with an orange flush on the side of the head, a dark patch on the chest and some developing bright yellow chests, front of arms and spots on dorsal surface of body. Femoral and preanal pores 40–49.

KEY CHARACTERS: Differs from the Painted Dragon (*C. pictus*) in having heterogeneous body scales (versus homogeneous). Differs from the Lake Disappointment Dragon (*C. nguyarna*) in having enlarged scales tending to form transverse bars on the back near midline (versus enlarged paler scales tending to form transverse bars on flanks).

DISTRIBUTION AND ECOLOGY: Semiarid to arid southern interior of WA, east to Serpentine Lakes, SA, mainly in chenopod shrublands along the sandy margins of claypans and salt lakes. Burrows are excavated at the bases of vegetation, often in raised slopes of sand consolidated by the root systems.

BIOLOGY: Like its close relative the Painted Dragon, this species' burrows are generally U-shaped with an escape exit terminating near the surface. They are mainly terrestrial, resting only on very low perches. A clutch of 3 eggs is recorded, with laying occurring during September and hatchlings sighted at the same locality (Newman Rocks, WA) during mid-February (Chapman & Dell, 1985; Greer, 1989). Hatchlings probably take more than one year, probably two, to achieve sexual maturity.

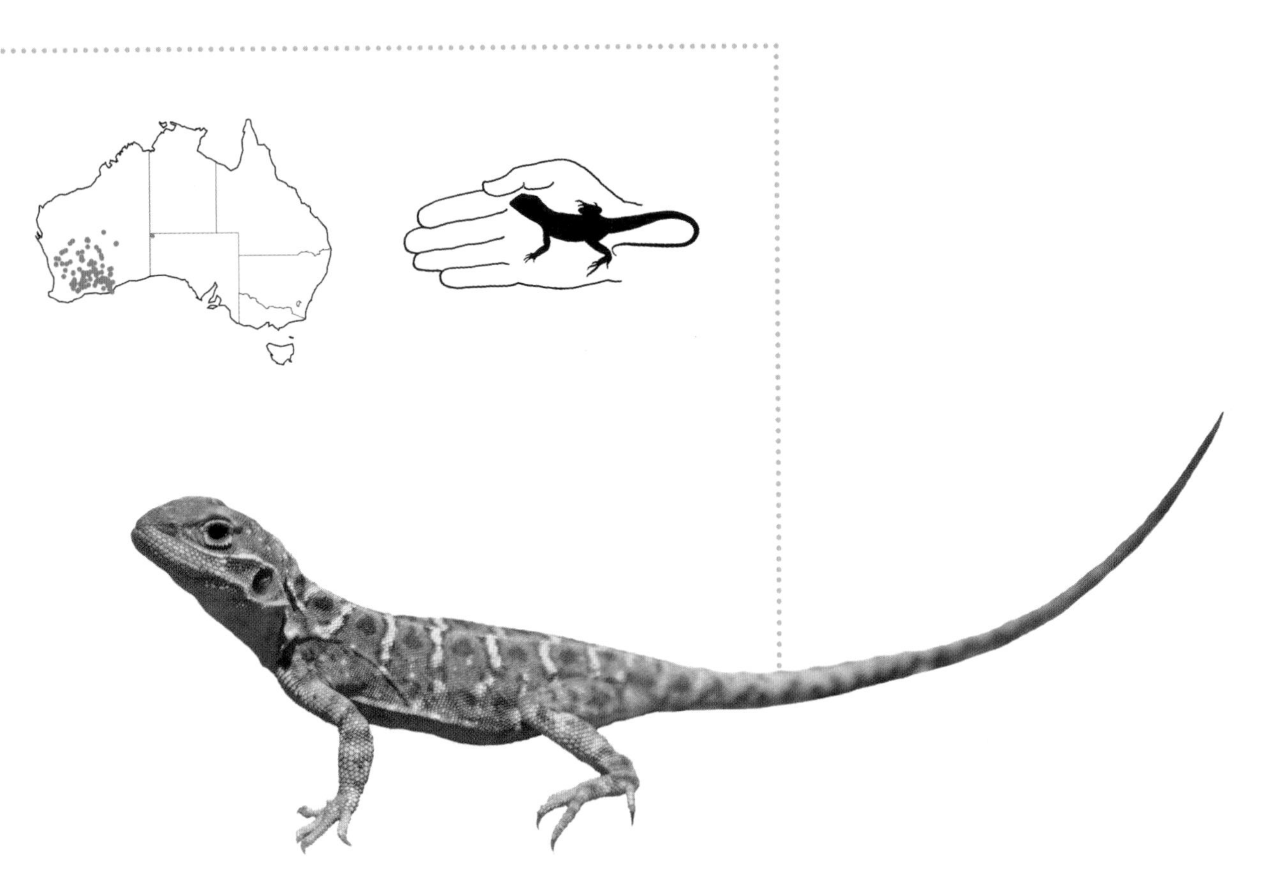

this page
*Ctenophorus salinarum*. Female.
Menzies area, WA
photo J. Melville

opposite page
right:
*Ctenophorus salinarum*. Male.
Menzies area, WA
photo R. Glor

left:
*Ctenophorus salinarum* habitat.
Lake Ballard, WA
photo J. Melville

## LOZENGE-MARKED DRAGON

*Ctenophorus scutulatus* Stirling and Zeitz, 1893

DESCRIPTION: SVL 115 mm. Head moderately long and narrow, body slender and limbs and tail relatively long. Body scales homogeneous: small and uniform with nuchal crest of compressed triangular scales continuous with a keeled vertebral line to base of tail, and dorsolateral spines on neck. Pale reddish brown to greyish brown with a series of short dark angular bars across nape and forebody, and a broad pale dorsolateral stripe which is constricted to form a series of dark-centred lozenge-shaped blotches between nape and base of tail. Throat marked with grey wavy lines. On breeding males there is a broad black stripe on throat and a black patch on chest extending to forearms and abdomen. A total of 39 - 60 femoral and preanal pores extend along full length of thighs.

KEY CHARACTERS: Differs from the Crested Dragon (*C. cristatus*) in lacking dorsolateral crests and from McKenzie's Dragon (*C. mckenziei*) in having paler colouration and attaining larger size (SVL 115 mm versus 76 mm).

DISTRIBUTION AND ECOLOGY: Semi-arid woodlands of the south-western interior, favouring, hard to stony soils supporting acacia and eucalypt woodlands and chenopod shrublands. Extremely swift and terrestrial, perching on the ground or low timber and dashing on hind limbs with forelimbs held to chest if disturbed. Although this species has been reported to use burrows (Greer, 1989), more recent work has suggested that *C. scutulatus* uses hollow logs as retreats or sleeps in the open (Thompson & Withers, 2005). However, has been observed to flee into a burrow when disturbed (JM pers.obs.).

BIOLOGY: Ants and termites are significant components of a broad arthropod diet. Clutches of 5–10 eggs are recorded (Pianka, 1971b).

opposite page right:
*Ctenophorus scutulatus.*
Charles Darwin Nature Reserve, WA
photo S. K. Wilson

opposite page left:
*Ctenophorus scutulatus* habitat.
Charles Darwin Nature Reserve, WA
photo S. K. Wilson

below:
*Ctenophorus scutulatus.*
Charles Darwin Nature Reserve, WA
photo S. K. Wilson

*Ctenophorus scutulatus.*
Lake Yeo, WA
photo J. Melville

## SLATER'S RING-TAILED DRAGON

*Ctenophorus slateri* Storr, 1967

**DESCRIPTION:** SVL 100 mm. Head and body relatively deep and robust, with long limbs and tail. Body scales mostly homogeneous; overlapping with spinose keels (keels are black in central Australia) converging back towards midline, largest on back and becoming much smaller and smooth to weakly keeled on flanks. Nuchal crest present, comprising small, raised, laterally compressed scales, and a raised vertebral ridge of aligned keels on slightly enlarged scales. Sexes differ. Mature males are dull fawn to reddish brown with weak pattern; dark paravertebral spots sometimes present, sometimes a reddish flush on nape, forebody and arms, a dark triangular patch on chest and obscure brown bands on tail. Females are dull reddish brown with widely separated paravertebral series of small dark dots. Pale dorsolateral dots are sometimes present, occasionally forming transverse rows. Populations occurring on the sandstone escarpments of Arnhem Land, NT, are pale grey, matching the rock colour, with strongly contrasting black and white bands on the last half of the tail. Males in these populations have a black chest and orange running along either side of the posterior half of the belly. Femoral and preanal pores 27–39.

**KEY CHARACTERS:** Central Australian population differs from the geographically close Graaf's Ring-tailed Dragon (*C. graafi*) in having sharp, black keels on dorsal scales.

**DISTRIBUTION AND ECOLOGY:** A rock specialist, occurring on a variety of ranges and outcrops across northern and central Australia, between the Kimberley region of WA and the interior of Qld. Population on the sandstone escarpments of Arnhem Land, NT, is associated with expansive rock shelves, with large flat exfoliations providing shelter (Sadlier, 1990). Lizards generally select protruding rocks as elevated vantage points and retreat to crevices if threatened.

**BIOLOGY:** Poorly known but believed to be similar to the closely related Western Ring-tailed Dragon (*C. caudicinctus*).

**NOTES:** Recent genetic studies (Melville *et al*, 2016) have elevated this species from its previous placement as a subspecies of *Ctenophorus caudicinctus*.

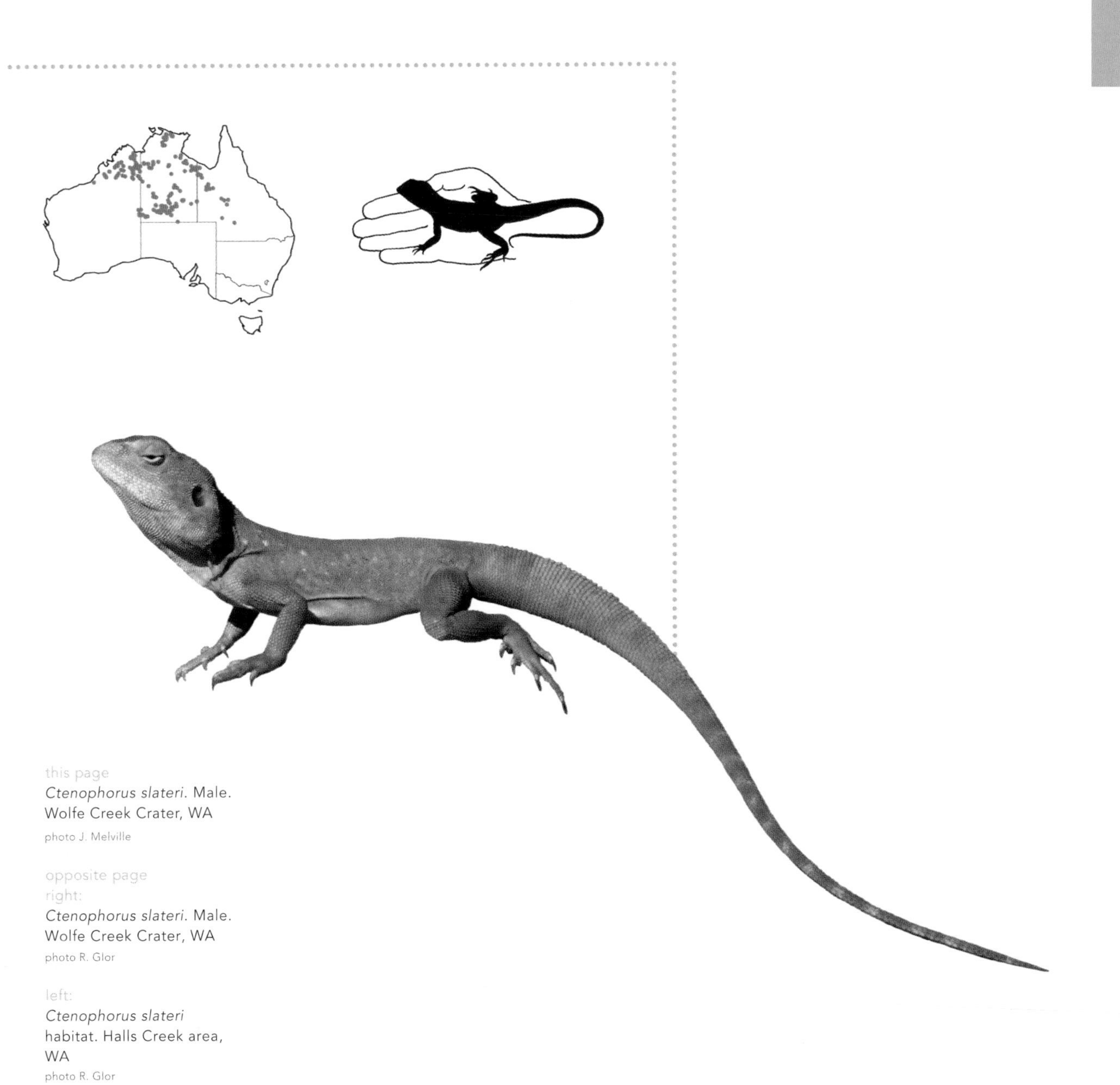

this page
*Ctenophorus slateri*. Male. Wolfe Creek Crater, WA
photo J. Melville

opposite page
right:
*Ctenophorus slateri*. Male. Wolfe Creek Crater, WA
photo R. Glor

left:
*Ctenophorus slateri* habitat. Halls Creek area, WA
photo R. Glor

top
*Ctenophorus slateri*. Female.
Kings Canyon
National Park, NT
photo J. Melville

bottom:
*Ctenophorus slateri*.
Morney Station, Qld
photo S. K. Wilson

opposite page:
*Ctenophorus slateri*. Male.
Western Arnhem Land, NT
photo R. Glor

## OCHRE DRAGON

*Ctenophorus tjantjalka* Johnston, *1992*

**DESCRIPTION:** SVL 73 mm. Head and body only moderately dorsally depressed. Scales on snout wrinkled to smooth but not keeled, and scalation on body homogeneous, with a vertebral keel line on back of males, and a low crest of enlarged, strongly keeled scales on nape. Sexes differ. Mature males are dark brown to grey. They may be patternless or they may have dark upper flanks and irregular cream vertical bars which darken to salmon pink on the flanks. The vertebral region is brown to grey and the tail has obscure bands. Females and juveniles are pale greyish brown to brick red with dark flecks and transverse rows of pale spots. Femoral and preanal pores 33–46.

**KEY CHARACTERS:** Differs from the Red-barred Dragon (*C. vadnappa*) in having pale bars or rings on tail (versus tail patternless). Males differ further in having flank pattern (when present) comprising cream to pink bars against a greyish background (versus red against a black background).

**DISTRIBUTION AND ECOLOGY:** Arid interior of SA, from Painted Hill area north to Oodnadatta and west to the base of Everard Ranges. Inhabits low, exposed and crumbling outcrops and gibber-strewn hills.

**BIOLOGY:** Unlike its rock-inhabiting relatives, this species does not occur on any substantial outcroppings, preferring more weathered stony terrain. Males of its closest relatives, the Red-barred Dragon (*C. vadnappa*), Tawny Dragon (*C. decresii*) and Peninsular Dragon (*C. fionni*), are all well-known for their impressive displays to females and rivals but this has not yet been reported for the poorly known Ochre Dragon.

this page
*Ctenophorus tjantjalka*. Female.
80km north of Coober Pedy, SA
photo S. Macdonald

opposite page
right:
*Ctenophorus tjantjalka*. Male.
Mt Willoughby, SA
photo A. Fenner

left:
*Ctenophorus tjantjalka* habitat.
80km north of Coober Pedy, SA
photo M. Clancy

## RED-BARRED DRAGON

*Ctenophorus vadnappa* Houston, 1974

DESCRIPTION: SVL 85 mm. Head and body moderately dorsally depressed. Scales on snout wrinkled to smooth but not keeled, and scalation on body homogeneous, with a vertebral keel line on back and a low crest of enlarged, strongly keeled scales on nape. Colouration differs markedly between the sexes. Males have a grey to bright blue vertebral stripe and black flanks with vertical rows of bright red blotches. Throats are striped with pale blue, and flushed with bright yellow over the gular fold area. Females and juveniles are brown with vertical rows of dark flecks. Femoral and preanal pores 34–46.

KEY CHARACTERS: Differs from the Tawny Dragon (*C. decresii*) and the Peninsula Dragon (*C. fionni*) in having wrinkled (versus smooth to keeled) scales on the snout and a vertebral keel line on back. Differs from the Ochre Dragon (*C. tjantjalka*) in having the tail uniform (versus with pale bars or rings) and flanks black with red bars (versus pattern absent, or grey with cream to pink markings).

DISTRIBUTION AND ECOLOGY: Rock inhabiting, on the northern Flinders Ranges and on scattered hills and outcrops north of Lake Torrens, SA. The dragons perch on elevated sites, and if disturbed dash under a rock or into a crevice.

BIOLOGY: This spectacular dragon is a stunning sight, basking on rocky scree slopes. The more brightly coloured males are at greater risk of predation, indicating there is a trade-off between predation risk and reproductive success (Stuart-Fox *et al*, 2003). The territorial males perform one of the most complex and impressive displays among Australian dragons. They align themselves parallel to each other, often facing opposite directions, lower their brightly coloured throats, raise their nuchal crests and vertebral ridges, elevate and lower their bodies by raising their hind limbs, and coil their tails vertically (Gibbons, 1979).

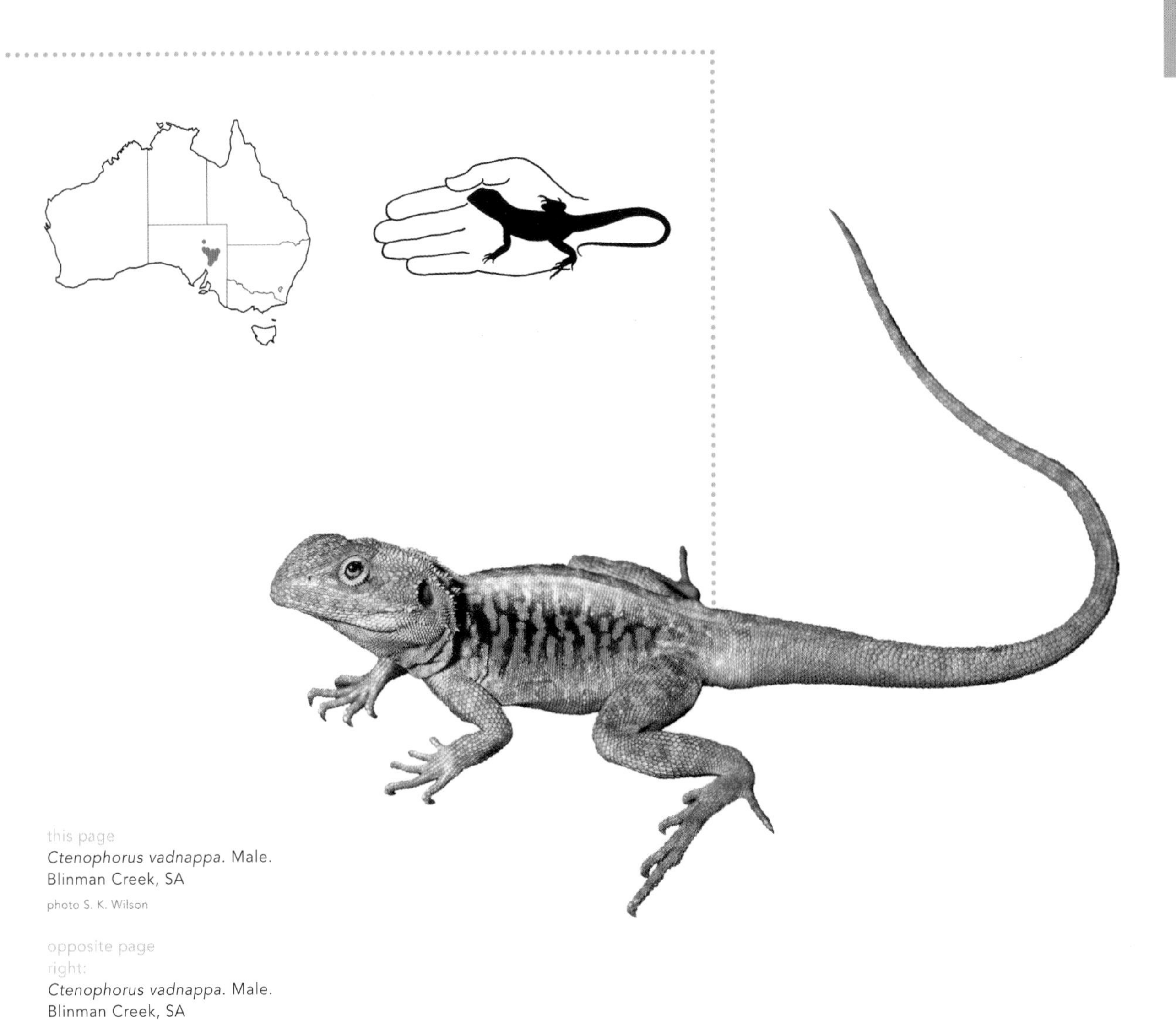

this page
*Ctenophorus vadnappa*. Male.
Blinman Creek, SA
photo S. K. Wilson

opposite page
right:
*Ctenophorus vadnappa*. Male.
Blinman Creek, SA
photo S. K. Wilson

left:
*Ctenophorus vadnappa*
habitat. Flinders Ranges, SA
photo S. K. Wilson

*Ctenophorus vadnappa*. Female.
Gammon Ranges, SA
photo S. K. Wilson

## YINNIETHARRA ROCK DRAGON

*Ctenophorus yinnietharra* Storr, 1981

**DESCRIPTION:** SVL 87 mm. Head and body dorsally depressed. Scalation on body is homogeneous; small uniform scales, each with a low keel converging towards midline. Enlarged spiny scales present below and behind ear and on sides of neck. Colouration varies between sexes. Adult males are bluish grey with obscure dark grey marbling, a reddish brown flush on midline and a dark streak from below eye to ear. The distal two thirds of the tail is boldly ringed with black and orange-brown. Females are dull reddish brown to greyish brown with dark grey hollow blotches, and often a weak indication of narrow irregular pale vertebral and transverse dorsal lines, and narrow irregular bands on proximal third of tail. Juveniles are similar, but with four broad dark bands on distal two thirds of tail. Femoral and preanal pores 40–50.

**KEY CHARACTERS:** The dorsally depressed body distinguishes this it from all sympatric species. Differs further from the Western Ring-tailed Dragon (*C. caudicinctus*) in having well-developed spines below and behind ear.

**DISTRIBUTION AND ECOLOGY:** Known only from low weathered granite outcrops and associated gibber flats with sparse *Acacia* shrubs in vicinity of Yinnetharra Station, in arid central-western interior of WA. Basks on low rocks and shrubs, with several individuals occasionally basking together. Extremely swift, sprinting for considerable distances on all four limbs. Shelters in crevices, burrows or hollow timber.

**BIOLOGY:** No studies have been undertaken on this species. Males have been recorded to lash their conspicuous tails in the presence of females or other males. It occurs in close proximity to another rock-inhabiting species, the Western Ring-tailed Dragon (*C. caudicinctus*). However, from limited observations, the Yinnietharra Rock Dragon appears to occupy the more weathered, lower granite outcrops.

**NOTES:** Currently listed at a vulnerable species in WA.

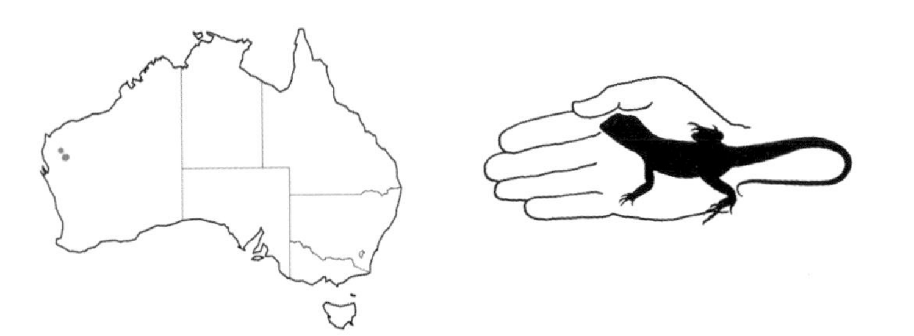

opposite page left:
*Ctenophorus yinnietharra* habitat.
Minnie Creek Station, WA
photo P. Cullen

opposite page right:
*Ctenophorus yinnietharra*. Female.
Yinnetharra Station, WA
photo S. K. Wilson

below:
*Ctenophorus yinnietharra*. Male.
Yinnetharra Station, WA
photo S. K. Wilson

*Diporiphora magna*.
Top Springs, NT
photo R. Glor

# DIPORIPHORA

## TWO-LINED DRAGONS

Genus *Diporiphora* Gray 1842

**DESCRIPTION:** Large and diverse genus, with 28 currently described species of small to moderate-sized dragons (SVL 34–94mm). Diagnostic features include: whether dorsal scales are homogeneous or heterogeneous (heterogeneous scalation ranges from enlarged paravertebral, dorsal, and/or dorsolateral rows, to 5 crests of enlarged spinose scales); whether keels on dorsal scales are parallel or converge towards midline; absence, presence and strength of gular, postauricular and scapular folds; whether enlarged canine teeth on the upper jaw are single or paired; and aspects of colour and pattern. Though highly variable between and within species, common elements of pattern often include a pair of pale dorsolateral stripes, dark transverse dorsal bars, a black circular blotch on shoulders and a pink to mauve flush on hips and tail-base.

**KEY CHARACTERS:** The genus can be very broadly viewed as comprising 5 evolutionary groups, including 3 species-rich groups and a pair of divergent, out-lying species, which have been identified based on genetic research.

The 2 species sitting outside the 3 species-rich groups are both unique in appearance and behaviour. They are not closely related to each other. The Mulga Dragon (*D. amphiboluroides*) has cryptic colouration, patterned with bark-like streaks to resemble the surrounding mulga. It is a slow-moving dragon that relies strongly on its excellent camouflage, and is restricted to the western interior and mid-west coast of WA. The Superb Dragon (*D. superba*) has an extremely long and slender body, very long thin limbs and digits, and an extremely long slender tail. Its body-shape and bright green colour distinguish it from all other Australian dragons. It is an arboreal species occurring along the edges of sandstone gorges in north-western Kimberley region, WA.

An eastern Australian group, extending from south-eastern Australia to eastern Cape York Peninsula, consists of the Common and Black-throated Nobbi Dragons (*D. nobbi* and *D. phaeospinosa*), Tommy Round-head (*D. australis*) and Black-throated Two-pored Dragon (*D. jugularis*). These are moderately to strongly robust dragons with heterogeneous dorsal scales, and adult males often have dark or black sides of the body, extending onto either side of neck. In *D. phaeospinosa* and *D. jugularis* this can be extensive, forming black throats. They inhabit a variety of forests, woodlands and mallees, well-timbered sandstone uplands and tropical savannah woodlands.

A group restricted to the northern monsoon tropics from Broome, WA, to western Qld, consists of the White-lipped Two-lined Dragon (*D. albilabris*), Northern Two-pored Rock Dragon (*D. sobria*), Robust Dragon (*D. bennettii*), and the Kimberley Two-pored Rock Dragon (*D. perplexa*). These are moderately to strongly robust dragons with a gular fold, usually having femoral pores and two (double) canine teeth on each side of the upper jaw. *D. bennettii* and *D. perplexa* are both rock-dwelling specialists, while *D. albilabris* and *D. sobria* are widely distributed generalists, occurring in tropical savannah woodlands.

*Diporiphora albilabris*.
Drysdale River Station, WA
photo S. K. Wilson

The remaining species, comprising the bulk of the genus, form a large and diverse evolutionary lineage. They are small and generally smooth, with no enlarged spinose crests (but paravertebral and dorsolateral scale rows may be enlarged). Most are slender, often similar in appearance, and have been subject to considerable confusion in identification. This group is most diverse in the savannahs and *Triodia*-dominated habitats across northern Australia but several species penetrate well into central and southern deserts.

**DISTRIBUTION AND ECOLOGY:** Most *Diporiphora* are semi-arboreal, perching on low timber, surface objects such as termite mounds and rocks, and on vegetation such as shrubs and *Triodia* hummocks. The Superb Dragon (*D. superba*) is a fully arboreal foliage inhabitant. Though alert and quick to fee if approached they are generally less swift than other similarly-sized dragons, tending to scuttle rather than sprint. The common elements of their dorsal patterns—pale stripes and dark bands—are distinctive features of lizards spanning several families world-wide, where the habitat exhibits a mix of pale grasses, dry twigs and stems. It appears well designed to conceal them against a complex backdrop.

**BIOLOGY:** *Diporiphora* have been poorly studied, and the identification of many species has been extremely problematic. The single or paired canine teeth, which are useful taxonomic features, tend to be much larger in males. Their function is not known but they may be employed during mating or perhaps male combat. Apart from the individual species descriptions, the only research conducted at the generic level have been focussed on WA and NT (Storr, 1974, and Doughty *et al*, 2012). More recently, Melville *et al* (2019) examined *Diporiphora* across northern Australia.

## CARNARVON DRAGON

*Diporiphora adductus* Doughty, Kealley & Melville, 2012

**DESCRIPTION:** SVL 60 mm. Very slender with long slender limbs and tail. No gular and postauricular folds, scapular fold absent or very weak. Dorsal scales homogeneous with keels parallel to midline. Ventral scales larger than dorsal scales. Dorsal colour light brown with pale silver vertebral stripe, prominent yellow-white dorsolateral stripes extending onto tail, and about 6 dark brown bands between dorsolateral stripes. Pale grey lateral stripe bordered by dark edge, commences in front of arm and continues posteriorly onto tail. Pale temporal stripe bordered by dark brown is continuous with dorsolateral stripes, but dark border finishes at ear. Ventral surface white with a pair of wide brown-grey stripes from snout, separating through gular region, converging on neck, then widening and separating on body before converging near hindlimbs and continuing onto tail as a single stripe. Femoral pores 0; preanal pores 0–2.

**KEY CHARACTERS:** The distribution of *D. adductus* does not overlap with other *Diporiphora* species but can be distinguished from nearby Pilbara species by the following combination of characters: no gular and postauricular folds, very weak or absent scapular fold, homogeneous dorsal scales with keels parallel to mid-line, ventral scales larger than dorsal scales, no femoral pores, and colour not being yellow-green.

**DISTRIBUTION AND ECOLOGY:** Occurs on low shrubs and grasses, particularly *Triodia*, *Acacia* and Beach Spinifex (*Spinifex longifolius*), usually on sand dunes or red sandy/loamy soils. Restricted to the northern part of the Carnarvon Basin, WA.

**BIOLOGY:** A semi-arboreal species for which little known about biology.

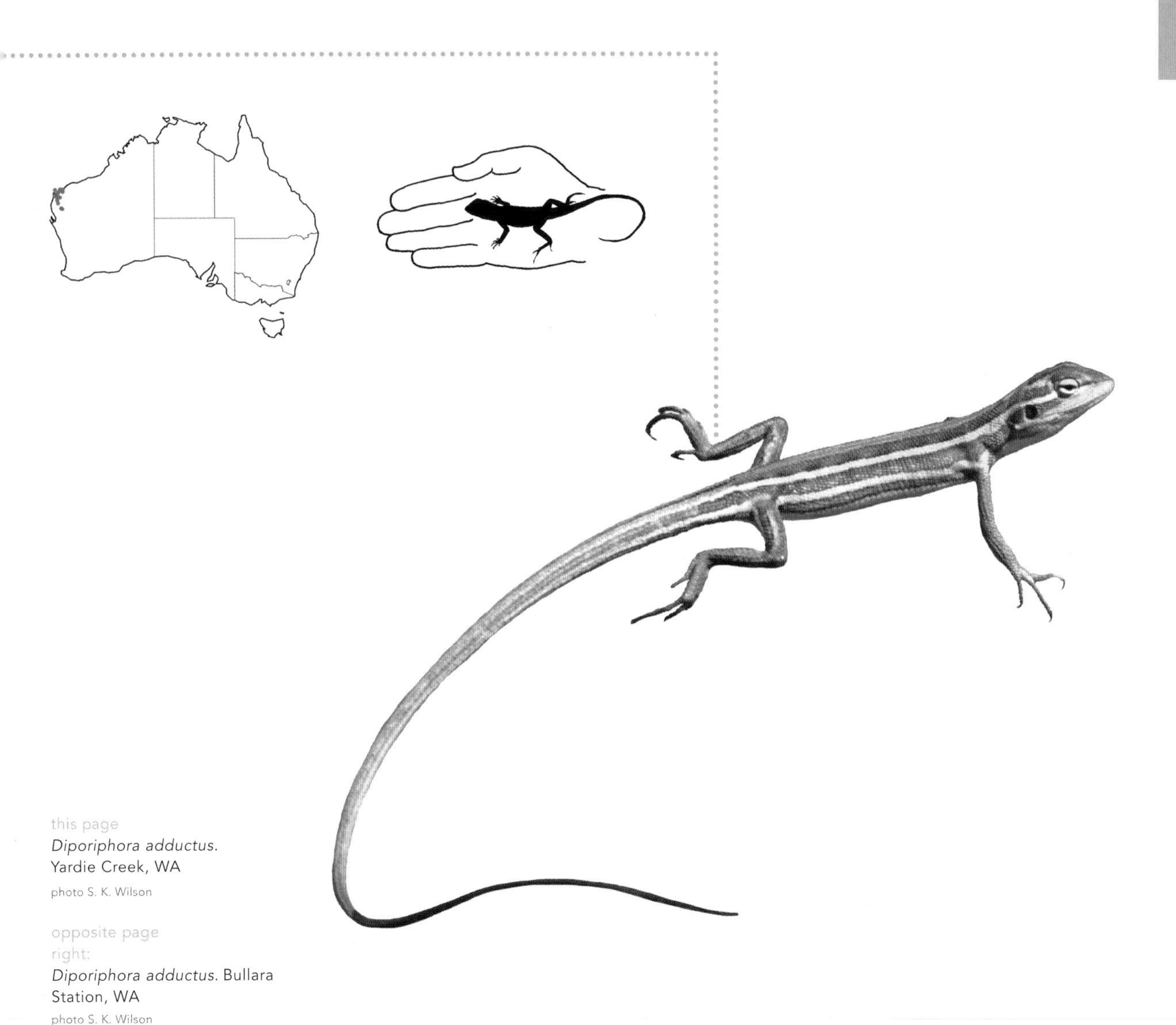

this page
*Diporiphora adductus.*
Yardie Creek, WA
photo S. K. Wilson

opposite page
right:
*Diporiphora adductus.* Bullara Station, WA
photo S. K. Wilson

left:
*Diporiphora adductus* habitat. Bullara Station, WA
photo S. K. Wilson

### WHITE-LIPPED TWO-LINED DRAGON

*Diporiphora albilabris* Storr, 1974

**DESCRIPTION:** SVL 55 mm. Moderately robust with long limbs and tail. Two canines on either side of upper jaw. Gular and postauricular folds present, scapular fold absent. Dorsolateral row of enlarged, strongly keeled white scales extend from back of head to hips along well-defined white dorsolateral stripes. Scales between these dorsolateral rows are strongly heterogeneous, with paravertebral rows, either side of the vertebral scale row, being reduced in size in comparison to both adjacent scale rows. Scales on the outer sides of the white dorsolateral stripes relatively small and keeled, with keels angling toward ventral surface at midbody. Dorsal patterning variable and complex with 5–6 dark brown bands between white dorsolateral stripes, intersected by a poorly defined grey vertebral stripe, about 3 scales wide. These dark bands extend irregularly beyond dorsolateral stripes, fading into the lateral colouration of dark brown with light brown spots. On breeding males, dorsal pattern tends to disappear. Head and upper body have strongly contrasting charcoal black, white and chestnut or orange-red, with throat, chest and tail also flushed with orange-red in some individuals. Labial scales pale cream with a few darker flecks, extending back as a broad pale band along jaw through back of head to enlarged spinose scales on the postauricular fold. Throat often has grey longitudinal stripes ending at gular fold. Ventral surfaces of body and tail cream and unpatterned. Femoral pores 2; preanal pores 4.

**KEY CHARACTERS:** Distribution overlaps a number of other *Diporiphora* species in the Kimberley. Differs from *D. sobria* in having strongly heterogeneous scales between enlarged dorsolateral rows (versus homogenous between dorsolateral rows). Differs from *D. perplexa* and *D. bennettii* in having heterogeneous dorsal scales (versus homogenous) and two femoral pores (versus no femoral pores). Differs further from *D. perplexa* in lacking dark markings on the tympanum (versus a black spot on tympanum). Differs from *D. magna* and *D. gracilis* in having a gular fold (versus absent), femoral pores (versus absent), double canine teeth in upper jaw (versus single) and white labial scales. Differs from *D. margaretae* in having a gular fold and double canine teeth on each side of upper jaw (versus single).

**DISTRIBUTION AND ECOLOGY:** Restricted to the central and northern Kimberley region. A habitat generalist occurring in tropical savannah woodlands and grasslands.

**BIOLOGY:** Little known about biology. Seen to perch on low vegetation, rocks or termite mounds.

**NOTES:** Following a recent taxonomic revision, this species' distribution has significantly decreased and much of the distribution formerly attributed to it is now occupied by *D. sobria* (Melville *et al*, 2019).

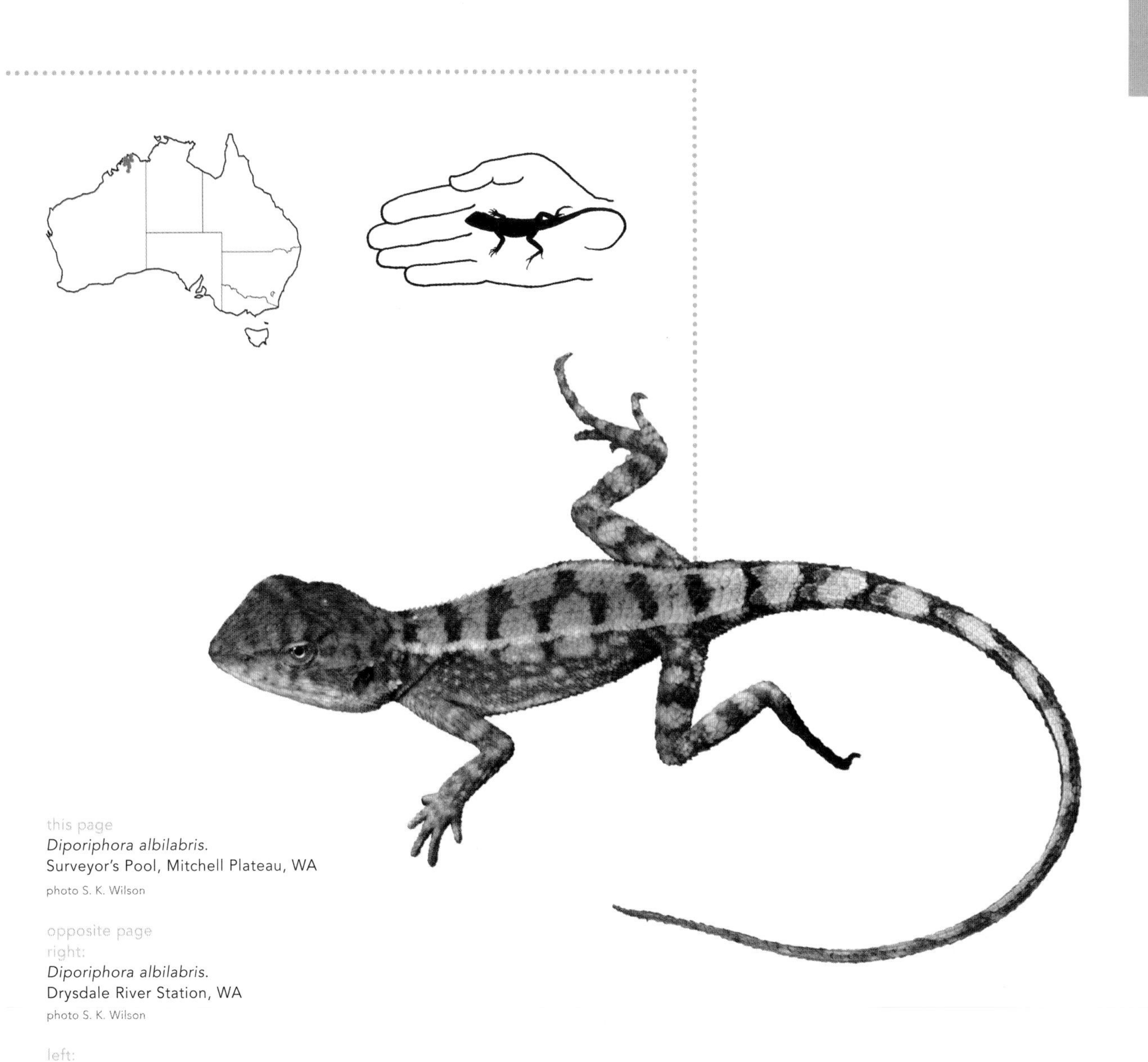

this page
*Diporiphora albilabris.*
Surveyor's Pool, Mitchell Plateau, WA
photo S. K. Wilson

opposite page
right:
*Diporiphora albilabris.*
Drysdale River Station, WA
photo S. K. Wilson

left:
*Diporiphora albilabris* habitat.
King Edward River area, WA
photo S. K. Wilson

*Diporiphora albilabris.*
Drysdale River Station, WA
photo S. K. Wilson

## AMELIA'S SPINIFEX DRAGON

*Diporiphora ameliae* Couper, Melville, Emmott & Chapple, 2012

**DESCRIPTION:** SVL 67 mm. Slender with long limbs and tail, A single canine on each side of upper jaw. No gular fold but with a scapular fold and a postauricular fold bearing small spines Dorsal scales homogeneous with keels parallel to midline and no crests. Dorsal colour mid-brown with pale grey vertebral stripe 2–3 scales wide, and prominent yellow dorsolateral stripes that continue onto tail; eight dark brown bands between dorsolateral stripes; pale lower lateral stripe usually present. Ventral pattern usually well-developed with three distinctive dark V-shaped markings on throat and gular, converging towards snout; the middle two extend back onto body, which has four longitudinal grey stripes on a cream background. Some individuals have little to no ventral pattern but all retain some indication of the throat pattern. Markings on both dorsal and ventral surfaces are most prominent in juveniles and become increasingly obscure in large adults, particularly large males. Breeding males have a prominent red flush on base of tail. Femoral pores 0; preanal pores 4.

**KEY CHARACTERS:** *D. ameliae* may overlap with *D. winneckei*, differing in having four continuous or broken ventral stripes, when present (versus only two stripes that are well spaced at midbody, converging to meet in the pectoral and pelvic regions). Additionally, *D. ameliae* has a strong spiny postauricular fold (versus absent).

**DISTRIBUTION AND ECOLOGY:** Mixed *Acacia aneura* and *A. ensifolia* woodland with spinifex on hard pebbly soils. Currently known from a restricted region between Bladensberg NP, Valetta and Noonbah Stations in western Qld. However, there are further suitable habitats in the Winton area and extending to the south.

**BIOLOGY:** Semi-arboreal species usually seen perching on spinifex.

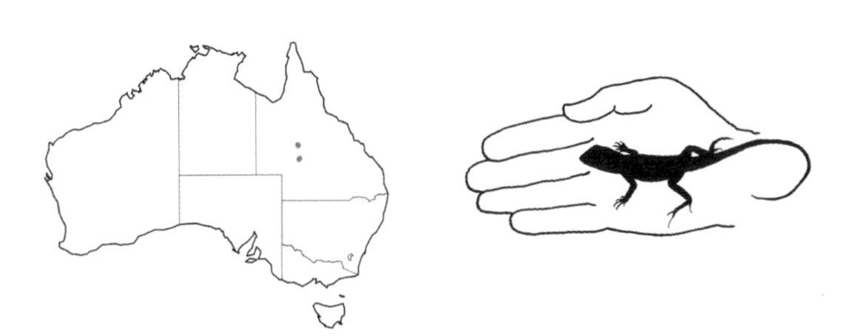

opposite page left:
*Diporiphora ameliae* habitat.
Valletta Station, Qld
photo S. Mahony

opposite page right:
*Diporiphora ameliae*. Breeding male.
Valetta Station, Qld
photo A. Emmott

below:
*Diporiphora ameliae*. Female.
Valetta Station, Qld
photo S. K. Wilson

## MULGA DRAGON

*Diporiphora amphiboluroides* Lucas and Frost, 1902

**DESCRIPTION:** SVL 94 mm. Short limbed with blunt-tipped tail and up-turned snout. Dorsal scales heterogeneous; small and weakly keeled with 5 enlarged dorsal crests, with the vertebral largest and continuous and the remainder small and disrupted. Grey with brown dash on snout, bat-winged markings between eyes, an oblique dash from above each eye towards nape and 3–5 longitudinal dashes along each side of back. Upper flanks dark grey. Sometimes obscure pale midlateral and ventrolateral stripes present. Femoral pores 0; preanal pores 1–3, aligned back towards midline.

**KEY CHARACTERS:** Differs from all sympatric species in having 5 enlarged dorsal crests.

**DISTRIBUTION AND ECOLOGY:** Arid to semiarid mulga woodlands and shrublands in the southern interior of WA, approaching the coast near Carnarvon. Generally associated with heavy, often stony soils. A semi-arboreal dragon that has been recorded at an average height above the ground of approximately 80cm (Pianka, 2013b), perching on the trunks and stems of low vegetation and among associated leaf litter, where the bark-like longitudinal streaked pattern provides excellent camouflage. Most specimens encountered serendipitously on roads.

**BIOLOGY:** Slow moving and easily overlooked, slowly sliding from direct view, rather than fleeing, if approached. Reported to feed mainly on termites (Ehmann, 1992). Average body temperature of 36.6°C has been reported and a clutch size of 12 (Pianka, 2013b).

this page
*Diporiphora amphiboluroides.*
Wilthorpe, WA
photo S. K. Wilson

opposite page
right:
*Diporiphora amphiboluroides.*
Mt Clare Station, WA
photo S. K. Wilson

left:
*Diporiphora amphiboluroides*
habitat. Yuin Station. WA
photo S. K. Wilson

## TOMMY ROUND-HEAD

*Diporiphora australis* Steindachner, 1867

**DESCRIPTION:** SVL 65 mm. Moderately robust with moderately long limbs and tail. Single canine on either side of upper jaw. Gular and postauricular folds present and scapular fold weak. Dorsal scales homogeneous but prominent keels form longitudinal ridges along back at the midline and a dorsolateral ridge on each side. These raised vertebral and dorsolateral scale rows extend forward to back of head and posteriorly onto tail. Scales in armpits small but not granular. Ventral scales strongly keeled. Patterning variable. Some individuals virtually lack pattern. A pair of pale dorsolateral stripes present or absent, with about 5 narrow broken pale cream bands across much broader dark background. Dark spot on sides of neck at anterior edge of shoulder. Flanks a similar colour to dorsal surface. Tail and rear legs strongly banded with light bands narrowest. Front legs weakly patterned. Ventral surface cream with no patterning. Femoral pores 0; preanal pores 4.

**KEY CHARACTERS:** Overlaps with *D. nobbi*, *D. phaeospinosa* and possibly *D. jugularis*. Differs from *D. nobbi* and *D. phaeospinosa* in being smaller and lacking spinose scales on thigh or neck, lacking dorsal crests (versus vertebral and dorsal rows of enlarged scales). Differs from *D. jugularis* in having a gular fold. Differs from *D. carpentariensis* in having a gular fold (versus absent) and lacking granular scales in armpit.

**DISTRIBUTION AND ECOLOGY:** Widespread along the northern east coast of Australia and adjacent inland regions, from southern Cape York to far northern NSW. A generalist species that occurs in dry open forests, woodlands and shrublands. A common sight in some of the northern cities. Appears to have adapted well to bushy suburban areas and has been found to be relatively tolerant to habitat disturbance (Kutt *et al*, 2011).

**BIOLOGY:** Although a common species in suburban areas in eastern Qld, relatively little is known about its biology. A territorial species, using low vegetation, fallen timber and termite mounds as perches.

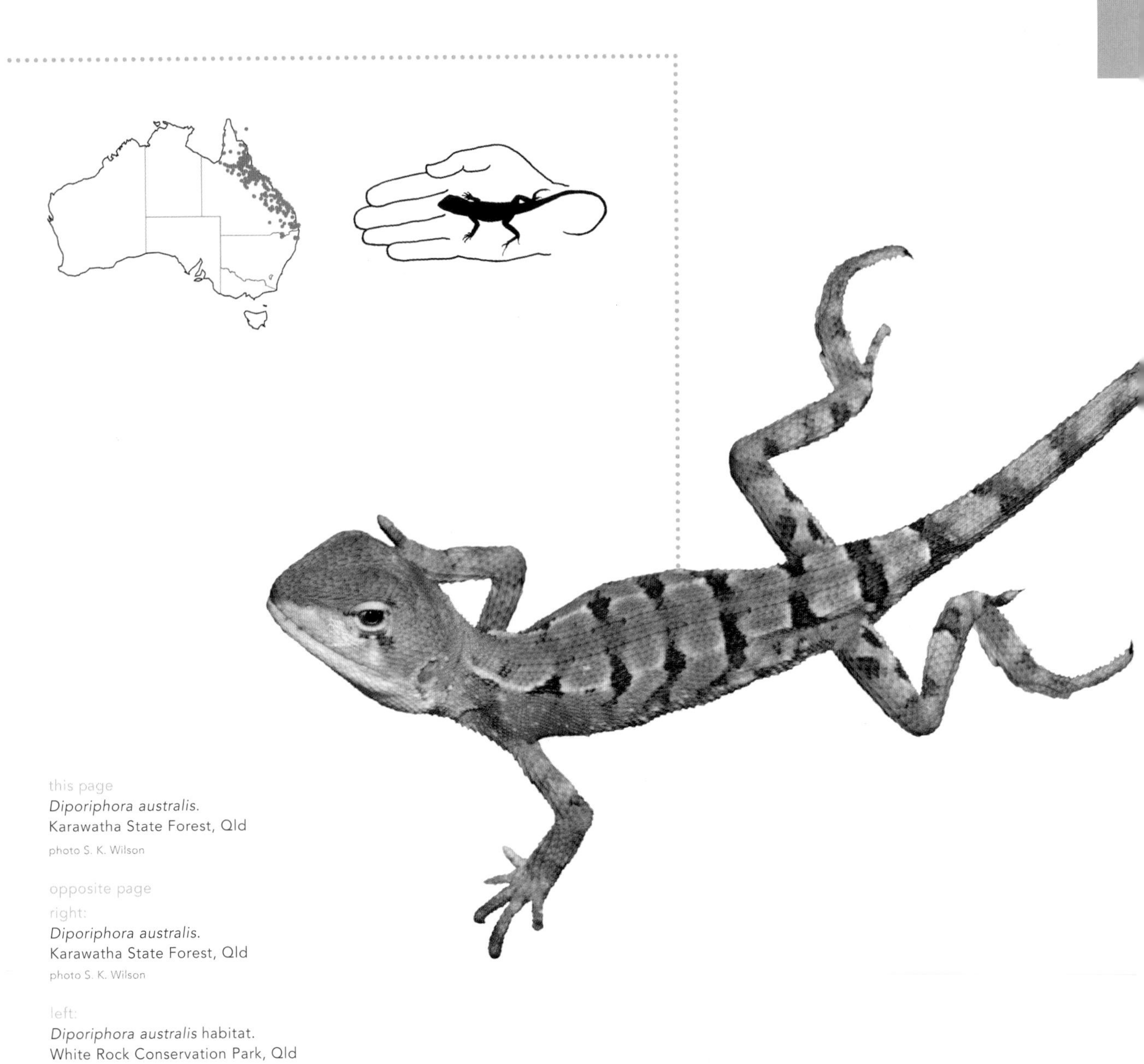

this page
*Diporiphora australis*.
Karawatha State Forest, Qld
photo S. K. Wilson

opposite page
right:
*Diporiphora australis*.
Karawatha State Forest, Qld
photo S. K. Wilson

left:
*Diporiphora australis* habitat.
White Rock Conservation Park, Qld
photo S. K. Wilson

above:
*Diporiphora australis.*
Thornborough, Qld
photo K. Chaplin

opposite page:
*Diporiphora australis.*
Bowling Green Bay National Park, Qld
photo S. K. Wilson

## ROBUST TWO-LINED DRAGON

*Diporiphora bennettii* Gray, 1845

**DESCRIPTION:** SSVL 45 mm. Small and robust with wide head, short thick neck, short limbs and short tail—no more than twice as long as SVL. A double-pair of enlarged canines on each side of upper jaw. Gular fold present, postauricular fold strong (usually with 1–3 spines) and scapular fold moderate to weak. Dorsal scales mostly homogeneous. No nuchal crest, dorsal keels parallel to midline and keels on flanks directed back to midline. Background colour greyish brown to reddish brown with dorsal pattern complex and diffuse. No prominent vertebral or dorsolateral stripes. Usually lacks dark spot on tympanum. Irregular brown blotches or transverse bands on dorsal surface with mixed light, medium, and dark scales, often forming short irregular lines. Ventral surface pale with occasional stippling on chin. Femoral pores 0; preanal pores 2–4.

**KEY CHARACTERS:** Similar to *D. perplexa*, with extensive distributional overlap and occurring in similar rocky habitats, but differs in having a short limbs and tail and lacking any dorsolateral stripes. Differs from *D. margaretae* and *D. gracilis* in having a gular fold and double canine teeth on each side of upper jaw. Can be distinguished from *D. albilabris* in lacking femoral pores, having a short tail, homogeneous (versus heterogeneous) dorsal scales, and lacking any prominent vertebral or dorsolateral stripes.

**DISTRIBUTION AND ECOLOGY:** Rock outcrops and escarpments on the Kimberley Plateau, occurring in the north-west Kimberley, including a few neighbouring offshore islands (Palmer *et al*, 2013).

**BIOLOGY:** A rock dwelling species for which little known about biology. Its true identity has been subject to some confusion, with other *Diporiphora* species mis-identified as *D. bennettii* in many publications.

this page
*Diporiphora bennettii.*
Little Merten's Falls, Mitchell Plateau, WA
photo S. K. Wilson

opposite page
right:
*Diporiphora bennettii.*
Little Merten's Falls, Mitchell Plateau, WA
photo S. K. Wilson

left:
*Diporiphora bennettii* habitat.
King Edward River area, WA
photo S. K. Wilson

### TWO-LINED DRAGON

*Diporiphora bilineata* Gray, 1842

**DESCRIPTION:** SVL 56 mm. Moderately gracile with long limbs and tail. Single canine on either side of upper jaw. Gular and postauricular folds absent. Scapular fold present. Dorsal scales heterogeneous; vertebral scale row, plus the 3–4 rows immediately adjacent are enlarged and strongly keeled. The vertebral and fourth longitudinal scale rows are raised. Beyond these enlarged scales are 4 rows of small homogeneous scales and then an enlarged, raised dorsolateral row, with those on each side strongly keeled. Granular scales in armpits, extending over arm onto neck. Ventral scales strongly keeled. Dorsal colour patterns variable, ranging from prominent and complex to virtually absent. Typically, a pair of narrow cream dorsal stripes overlie about 6 dark bands, significantly narrower than the pale interspaces, across back between neck and pelvis. Dark patch in armpit extending up onto shoulder. Flanks dark with dark granular scales extending posteriorly onto flanks, which are speckled with scattered white scales. Ventral surface cream with no patterning. Breeding males tend to lose some of their dorsal patterning, develop a yellow "wash" over head and upper body and a large back patch in armpit extending onto shoulder. Femoral pores 0; preanal pores usually 2.

**KEY CHARACTERS:** Overlaps with a few other *Diporiphora* species. Differs from *D. magna* in usually having 2 preanal pores (versus 4), lacking a postauricular fold (versus present), having heterogeneous dorsal scales comprising enlarged vertebral scales (versus homogeneous) and having dark flanks with scattered white scales. Differs from *D. sobria* in lacking femoral pores, lacking a gular fold and having single canines on either side of the upper jaw (versus double).

**DISTRIBUTION AND ECOLOGY:** A habitat generalist occurring in tropical savannah woodlands and grasslands of the northern NT. Occurs in fire prone habitats and has been found to be favoured by early-season hot fires (Braithwaite, 1987, Tainor & Woinarski, 1994). Contacts *D. magna* in the Pine Creek area.

**BIOLOGY:** Little known about biology. Seen to perch on low vegetation, rocks or termite mounds. Clutches of 4–8 eggs have been recorded (James & Shine, 1988).

**COMMENTS:** Recent taxonomic revision has contracted the distribution of this species to the northern portion of NT, with the population on Cape York, Qld, now identified as an unrelated species, *D. jugularis* (Melville *et al*, 2019). Previously, adult males of *D. bilineata* in breeding colouration may have been mistaken for *D. magna*. Genetic work has shown that these two species have only a narrow margin of overlap (Smith *et al*, 2011).

this page
*Diporiphora bilineata.*
Casuarina, NT
photo S. K. Wilson

opposite page
right:
*Diporiphora bilineata.*
Darwin, NT
photo R. Browne-Cooper

left:
*Diporiphora bilineata*
habitat. Kakadu National
Park, NT
photo R. Coupland

following page:
*Diporiphora bilineata.*
Douglas Daly region, NT
photo S. K. Wilson

## GULF TWO-LINED DRAGON

*Diporiphora carpentariensis* Melville, Smith Date, Horner & Doughty, 2019

**DESCRIPTION:** SVL 62 mm. Gracile with long limbs and tail. Single canine on either side of upper jaw. Gular fold absent. Strong to weak postauricular fold and strong scapular fold, extending slightly onto ventral surface on either side of gular region. Cluster of small spinose scales on postauricular fold with one cream-coloured spine clearly the largest. Homogeneous, strongly keeled dorsal and ventral scales. Granular scales in armpit, extending along full length of scapular fold. Scales on neck, anterior to scapular fold, small but not granular. Pattern variable, from strongly patterned to plain. Patterned individuals have pale dorsolateral stripes intersected by 8–9 dark brown bands, often offset on either side of a broad greyish vertebral stripe. Flanks flecked with a few light brown scattered scales. Plain individuals retain the pale dorsolateral stripes from neck to at least mid-back, and usually extending onto tail. Granular scales in axilla are dark brown or black, flanks cream, grey, or light brown with little patterning. No white markings on face, labial scales speckled with light brown flecks. Ventral surface cream, usually plain but sometimes faint dark flecking on throat. Breeding males, often with little dorsal patterning, have a large back patch in armpit extending onto shoulder and some have a pink flush on base of tail. Femoral pores 0; preanal pores 4–5 (usually 4).

**KEY CHARACTERS:** Distribution overlaps with a number of other *Diporiphora* species in the Gulf region of Qld. Very similar morphologically to *D. granulifera* and possibly contacting in the central Gulf region. Differs in having spinose scales on postauricular fold, with a single spine clearly largest (versus spinose scales on postauricular fold weak to absent), and lacking granular scales on sides of neck (versus extending anteriorly from scapular fold onto neck). Differs from *D. jugularis* in having a strong scapular fold (versus absent), having granular scales in armpit that extend over shoulder and along scapular fold (versus, scales in armpit not granular), having scales on flanks relatively homogeneous (versus strongly heterogeneous) and lacking a black gular band and/or black spot on sides of neck. Differs from *D. australis* in lacking a gular fold (versus present) and having granular scales in armpit that extend over shoulder and along scapular fold.

**DISTRIBUTION AND ECOLOGY:** A habitat generalist occupying tropical savannah woodlands and grasslands in far north-east Gulf region of Qld and central to western Cape York Peninsula.

**BIOLOGY:** Little known about biology. Seen to perch on low vegetation, rocks or termite mounds.

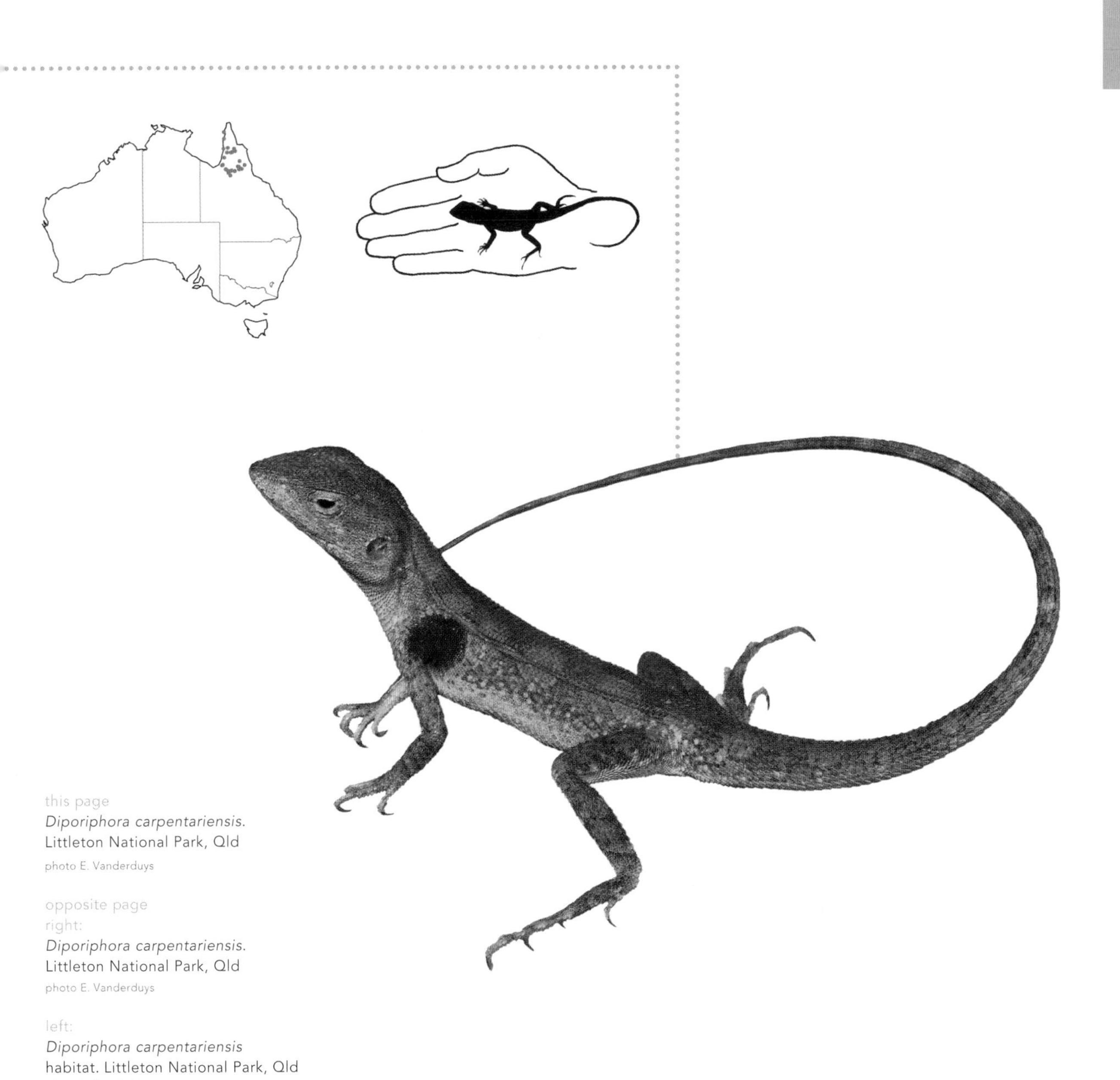

this page
*Diporiphora carpentariensis.*
Littleton National Park, Qld
photo E. Vanderduys

opposite page
right:
*Diporiphora carpentariensis.*
Littleton National Park, Qld
photo E. Vanderduys

left:
*Diporiphora carpentariensis*
habitat. Littleton National Park, Qld
photo E. Vanderduys

### CRYSTAL CREEK TWO-LINED DRAGON

*Diporiphora convergens* Storr, 1974

**DESCRIPTION:** SVL 34 mm. A small dragon with long limbs and tail. Gular and scapular folds very strong, postauricular fold absent. Homogeneous dorsal scales with keels converging to midline. No crests, a few small postauricular spines. Lateral scales a similar size to those on the dorsal surface and converging strongly towards the dorsum. Ventral scales weakly keeled. Dorsal surface has very little patterning, a few faint dark cross bars that are broken in the vertebral area. Femoral pores 0; preanal pores 0 (not discernible).

**KEY CHARACTERS:** Differs from all other *Diporiphora* in the Kimberley region in having dorsal scales with keels strongly converging towards the midline.

**DISTRIBUTION AND ECOLOGY:** Described from a single specimen collected in 1972 from Crystal Creek, in the north-west Kimberley region, WA. Ecology unknown.

**BIOLOGY:** Unknown.

**NOTES:** Only known from the single specimen on which the species was described. This specimen (possibly a sub-adult) is in relatively poor condition. No other dragons with converging keels on the dorsal scales have been located in the region.

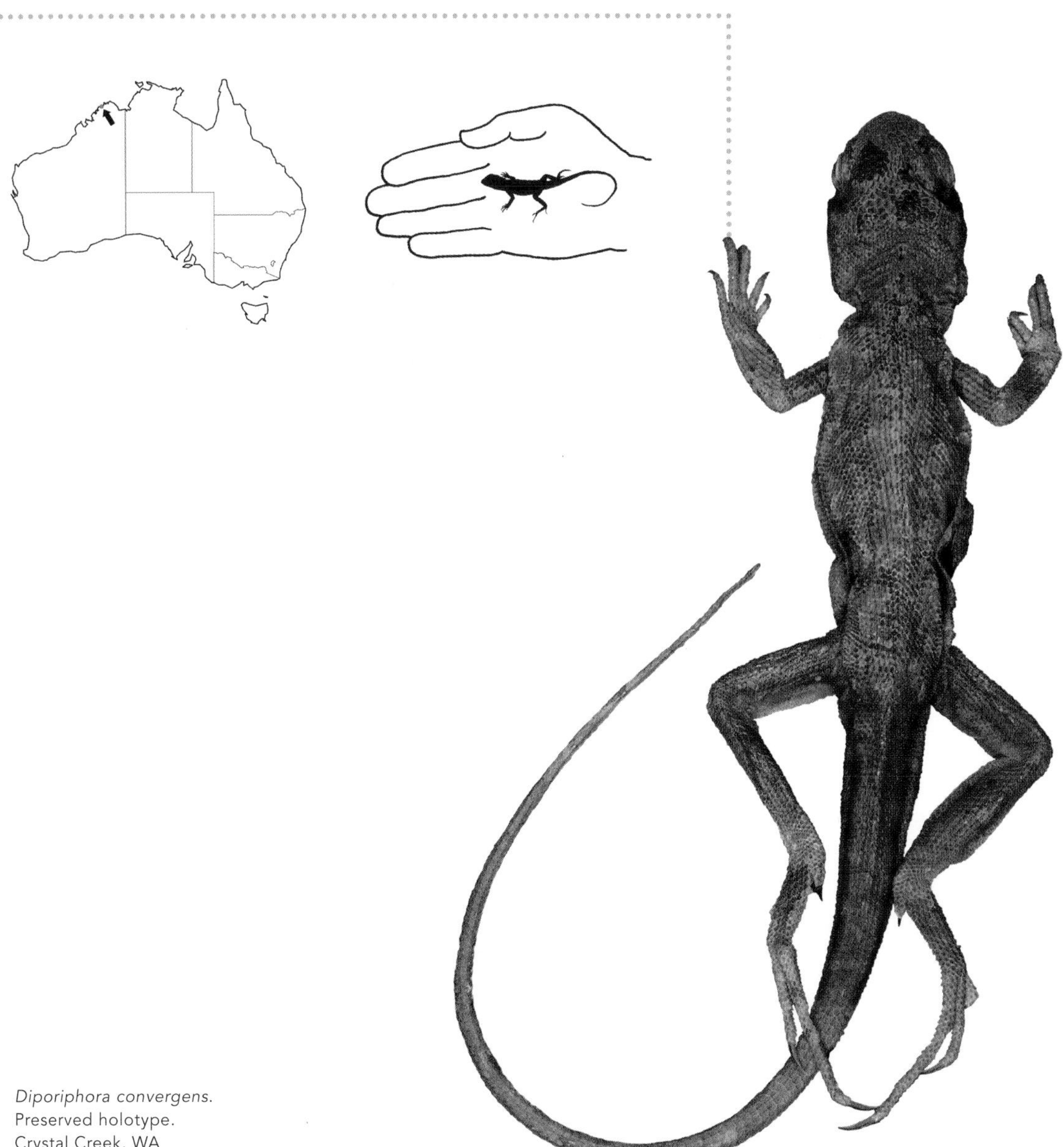

*Diporiphora convergens.*
Preserved holotype.
Crystal Creek, WA
photo B. Maryan

## GRACILE TWO-LINED DRAGON

*Diporiphora gracilis* Melville, Smith Date, Horner & Doughty, 2019

DESCRIPTION: SVL 57 mm. Gracile with long limbs and tail. Single canines on either side of upper jaw. Gular fold absent, scapular fold weak and postauricular fold weak or absent. Dorsal scales homogeneous and strongly keeled, with keels forming longitudinal ridges parallel to midline., Cluster of slightly enlarged spinose scales at rear of head. Granular scales in armpit, extending over arm to scapular fold. Ventral scales strongly keeled on body and weakly keeled on throat. Variable patterning from strongly patterned to plain individuals. Patterned individuals have pale dorsolateral stripes overlying dark brown bands offset on either side of a broad greyish vertebral stripe. No patterning on head and no pale stripe between eye and ear. Plain individuals have no dorsolateral or vertebral stripes, no pattern on head. The granular scales in armpit are dark brown, they have a well-defined black spot on sides and a greenish-yellow hue to body. Ventral surface cream, without markings. Breeding males have a pink flush on tail. Femoral pores 0; preanal pores 4.

KEY CHARACTERS: Distribution of *D. gracilis* overlaps numerous other *Diporiphora* species. Differs from *D. pindan* in lacking a well-defined white stripe between eye and ear (versus present) and having strongly keeled dorsal scales where keels form longitudinal ridges (versus more weakly keeled). Differs from *D. magna* in lacking strong postauricular and scapular folds, and having a more gracile build. Differs from *D. lalliae* in lacking gular fold (versus present) and having granular scales in armpit. Differs from *D. albilabris*, *D. sobria*, *D. perplexa* and *D. bennettii* in lacking femoral pores, lacking a gular fold and having single (versus double) canines on each side of the upper jaw.

DISTRIBUTION AND ECOLOGY: Restricted to the southwestern Kimberley region. Currently known from two locations, the type locality on the Fairfield-Leopold Downs Road and further east on Mornington Station. A specialist, occupying savannah grasslands on clay soils associated with the floodplain of the Lennard River.

BIOLOGY: Little known about biology. Has been observed perching in vegetation, including grasses and shrubs.

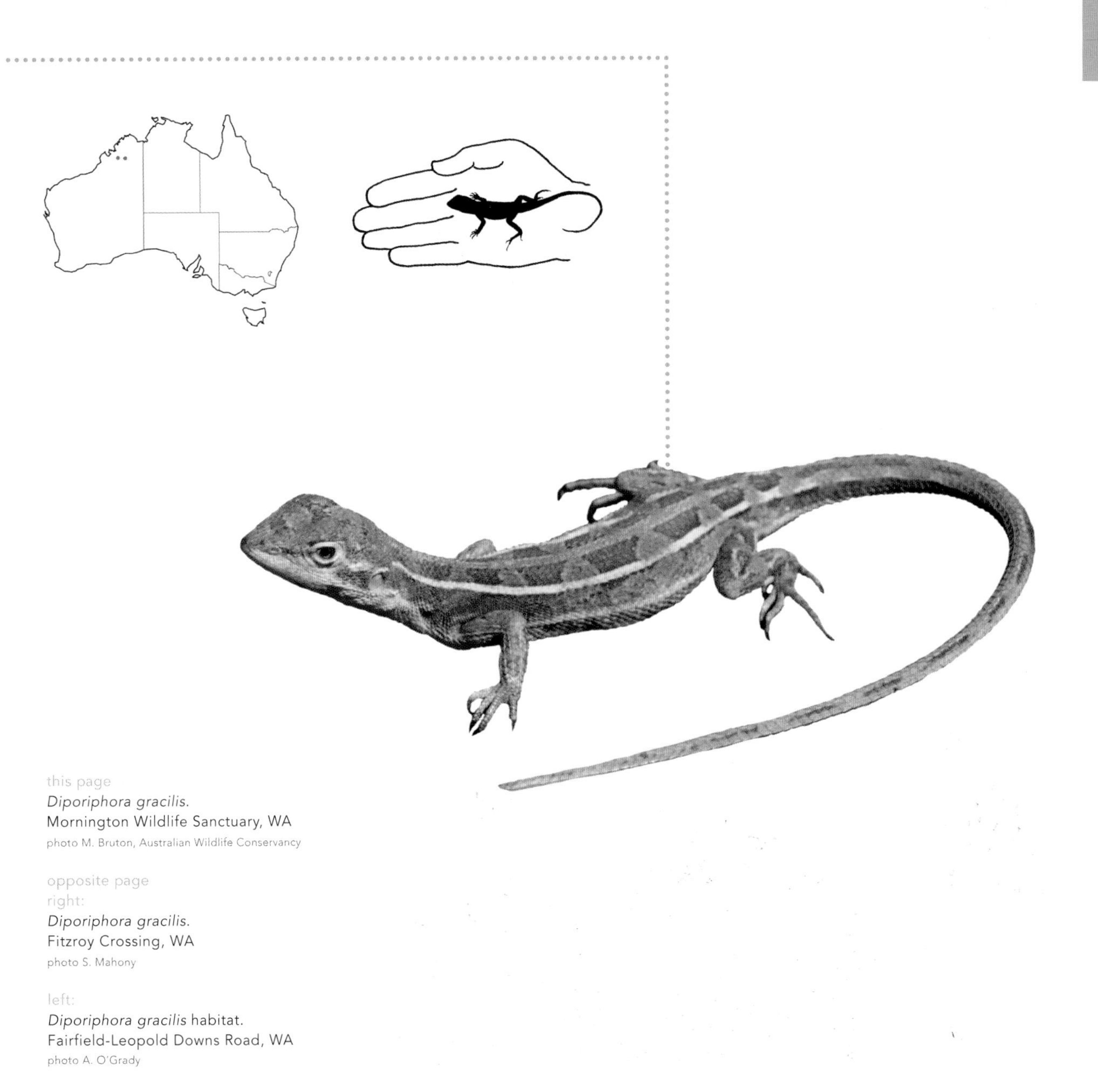

this page
*Diporiphora gracilis.*
Mornington Wildlife Sanctuary, WA
photo M. Bruton, Australian Wildlife Conservancy

opposite page
right:
*Diporiphora gracilis.*
Fitzroy Crossing, WA
photo S. Mahony

left:
*Diporiphora gracilis* habitat.
Fairfield-Leopold Downs Road, WA
photo A. O'Grady

this page
*Diporiphora gracilis.*
Mornington Wildlife Sanctuary, WA
photo M. Bruton, Australian Wildlife Conservancy

opposite page:
*Diporiphora gracilis.*
Mornington Wildlife Sanctuary, WA
photo M. Bruton, Australian Wildlife Conservancy

## GRANULATED TWO-LINED DRAGON

*Diporiphora granulifera* Melville, Smith Date, Horner & Doughty, 2019

**DESCRIPTION:** SVL 65 mm. Moderately gracile with long limbs and tail. Single canines on either side of upper jaw. Gular fold absent. Strong to weak postauricular fold with little or no indication of spinose scales, and strong scapular fold extending slightly onto ventral surface on either side of gular region. Homogeneous strongly keeled dorsal and ventral scales. Granular scales in armpit, extending over arm and along the full length of scapular fold. Small granular scales both in scapular fold and anteriorly onto side of neck. Variable patterning from strongly patterned to plain individuals. Patterned individuals have well-defined pale dorsolateral stripes overlying approximately 6–8 dark brown bands slightly offset to each other on either side of a broad undefined greyish vertebral stripe. Plain individuals have pale dorsolateral stripes. Granular scales in armpit are dark brown. No white markings on face, labial scales speckled with light brown flecks. Ventral surface cream, usually plain but some individuals have faint dark flecking on throat. Breeding males tend to have little patterning but have a large black patch in armpit extending onto shoulder, and some have a pink flush on base of tail. Femoral pores 0; preanal pores 4–6 (usually 4).

**KEY CHARACTERS:** Distribution contacts a number of other *Diporiphora* species in the western Gulf region. Very similar morphologically to *D. carpentariensis* and possibly overlapping in central Gulf region. *D. granulifera* differs in lacking or having little indication of spinose scales on postauricular fold (versus spines present including one clearly the largest) and having granular scales extending anteriorly from scapular fold onto neck (versus granular scales restricted to scapular fold). Differs from *D. magna* in having granular scales extending over shoulder, along scapular fold and onto ventral surface of neck (versus granular scales restricted to scapular fold). Differs from *D. lalliae* in lacking a gular fold. Can be distinguished from *D. sobria* in having homogeneous dorsal scales (versus raised dorsolateral row), lacking femoral pores and gular fold and having single (versus double) canines on either side of the upper jaw.

**DISTRIBUTION AND ECOLOGY:** Restricted to the far north-west Gulf region of Qld including the Northwest Highlands Bioregion, possibly extending west into NT. A habitat generalist occurring in tropical savannah woodlands and grasslands, particularly where spinifex is growing on rocky terrain.

**BIOLOGY:** Little known about biology. Has been observed perching on low vegetation, rocks or termite mounds and spinifex hummocks.

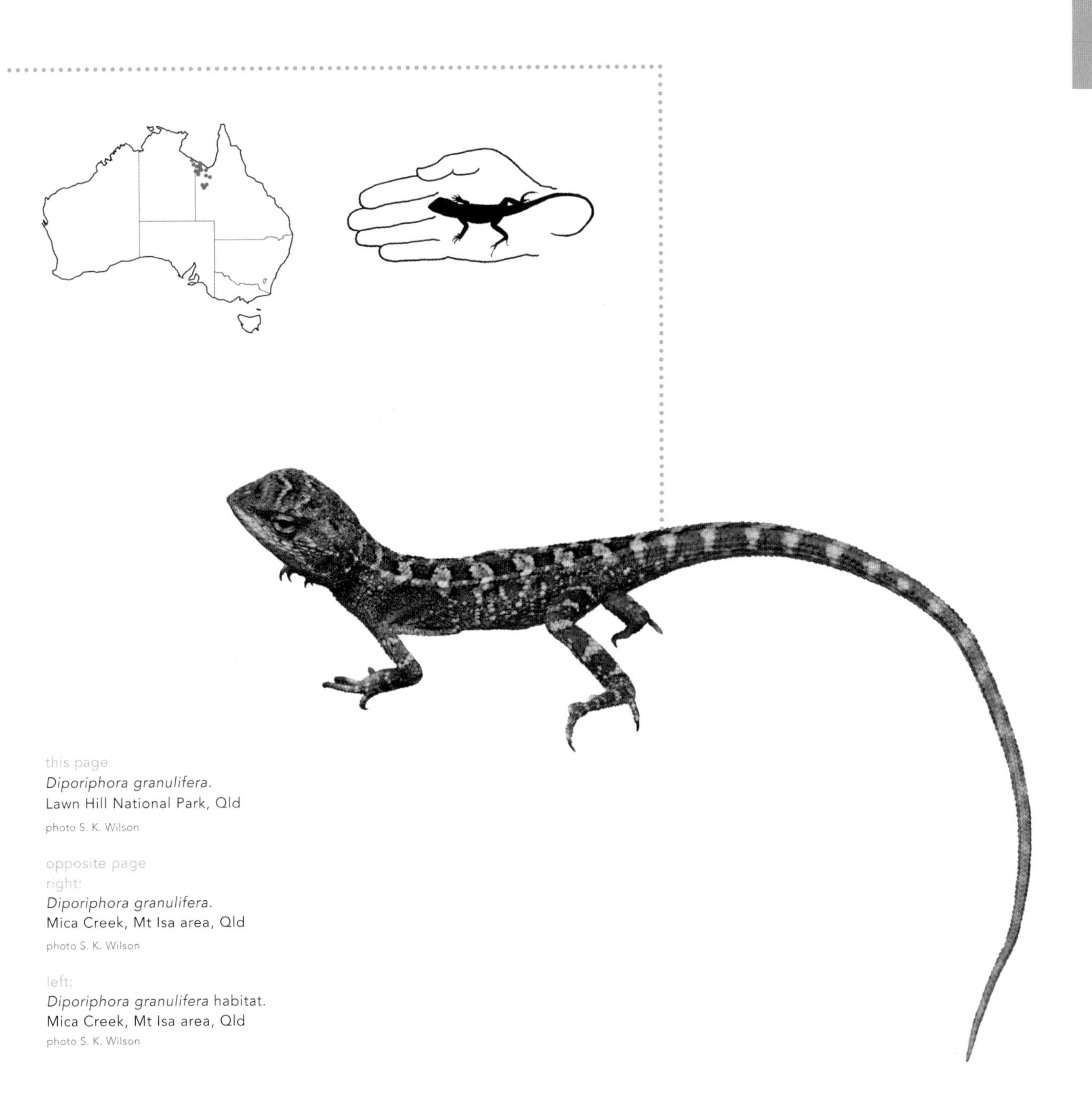

this page
*Diporiphora granulifera.*
Lawn Hill National Park, Qld
photo S. K. Wilson

opposite page
right:
*Diporiphora granulifera.*
Mica Creek, Mt Isa area, Qld
photo S. K. Wilson

left:
*Diporiphora granulifera* habitat.
Mica Creek, Mt Isa area, Qld
photo S. K. Wilson

*Diporiphora granulifera.*
Mt Isa, Qld
phot S. Wilson

*Diporiphora granulifera*.
Lawn Hill National Park, Qld
phot S. Wilson

## BLACK THROATED TWO-PORED DRAGON

*Diporiphora jugularis* Macleay, 1877

**DESCRIPTION:** SVL 68 mm. Moderately slender with long limbs and tail. Single canines on either side of upper jaw. Gular fold absent. Postauricular fold weak or absent and scapular fold absent. Dorsal scales homogeneous with prominent keels forming longitudinal ridges, including vertebral and dorsolateral ridges. Scales in armpits small but not granular. Scales on flanks strongly heterogeneous with scattered, distinctly larger scales. Ventral scales strongly keeled forming linear ridges. Patterning variable, from plain to strongly patterned individuals. Patterned individuals, often adult females and sub-adults, have approximately 4–6 broken dark brown bands across back on a pale background, broken by a pale grey or light brown vertebral stripe. Those with little patterning, often adult males, have pale cream dorsolateral stripes centred on strongly keeled dorsolateral ridges of scales. Flanks dark brown and the scattered enlarged scales are paler creating a speckled or flecked appearance. A broad dark brown or black band covering chin and throat is prominent in adult males, extending anteriorly to lips, and smaller or sometimes absent in females. Individuals missing the dark gular band may have a dark spot on either side of neck. Remaining ventral surfaces cream without patterning. Femoral pores 0; preanal pores 4.

**KEY CHARACTERS:** Overlaps with *D. carpentariensis*, *D. australis* and *D. nobbi*. Differs from *D. australis* and *D. nobbi* in lacking a gular fold, and further from *D. nobbi* in being smaller, lacking spinose scales on thigh or neck, lacking 5 dorsal crests and lacking femoral pores (versus 2–4 on each side). Differs from *D. carpentariensis* in lacking a scapular fold, scales in armpit are not granular, scales on flanks are strongly heterogeneous and has black gular band and/or black spot on sides of neck.

**DISTRIBUTION AND ECOLOGY:** Widespread along eastern and central Cape York Peninsula and islands of Torres Strait, Qld. Little is known about the ecology of this species but has been considered to be similar to *D. bilineata* in NT. It is a generalist species that occurs in dry open forests, woodlands and shrublands.

**BIOLOGY:** Little known about biology.

**NOTES:** *Diporiphora jugularis* has previously been referred to as *D. bilineata* on the basis of lacking a gular fold and most distributions maps for *D. bilineata* usually show a disjunct distribution, with a population on Cape York Peninsula in Qld. However genetic work has shown that this lineage is not closely related to *D. bilineata* and is instead the sister lineage to *D. australis* (Edwards & Melville, 2010). Recent taxonomic work has resurrected this population to species status (Melville *et al*, 2019).

this page
*Diporiphora jugularis.*
Mareeba wetlands, Qld
photo R. Coupland

opposite page
right:
*Diporiphora jugularis.* Mature male. Iron Range, Qld
photo S. K. Wilson

left:
*Diporiphora jugularis* habitat.
Lockhart River area, Qld
photo S. K. Wilson

## LALLY'S DRAGON

*Diporiphora lalliae* Storr, 1974

**DESCRIPTION:** SVL 62 mm. Moderately slender with long limbs and tail. Single canine on each side of upper jaw. Gular fold present and postauricular and scapular folds strong. Homogeneous, strongly keeled dorsal scales. Scales in armpit small but usually not granular. Dark spot in armpit absent but some individuals have dark spot over shoulder. Often a prominent white stripe running from behind eye to above the ear. Lacks dark spot on tympanum. Dorsal patterning varies from strongly to weakly patterned. Strongly patterned individuals have 5–6 wide, dark brown bands between shoulders and hips, intersected by a wide grey vertebral stripe, and pale dorsolateral stripes. At the shoulder these dorsolateral stripes consist of an enlarged longitudinal row of scales. More plain individuals usually have pale dorsolateral stripes from neck to at least mid-back but these are sometimes absent. Ventral usually un-patterned but sometimes with pair of dark parallel stripes. Breeding males often develop a pink flush on sides of tail. Femoral pores 0; preanal pores 4.

**KEY CHARACTERS:** Differs from *D. gracilis*, *D. granulifera* and *D. magna* in having a gular fold (versus absent). Differs from *D. sobria* in having homogeneous dorsal scales (versus heterogeneous; dorsolateral rows raised), single canine tooth on each side of upper jaw (versus 2 canine teeth on each side) and lacking femoral pores. Differs from *D. pindan* in having a gular fold and strong postauricular and scapular folds (versus gular and postauricular folds absent and scapular fold weak).

**DISTRIBUTION AND ECOLOGY:** Occurs in a variety of habitats from savannah woodlands and grasslands to arid habitats, usually with a *Triodia* understorey. They occur in a band across the northern extent of the arid zone from the Kimberley Region to western Qld, extending into the southern monsoon tropics,

**BIOLOGY:** A generalist species which occupies many habitats. Often seen perching on small rocks, termite mounds or clumps of earth.

**COMMENTS:** Until recently this species had been difficult to define, with lizards of several species being attributed to *D. lalliae*. Recent genetic work has assisted with identifying key characters and has narrowed the distribution to the northern arid zone and into the southern monsoon tropics (Melville *et al*, 2019).

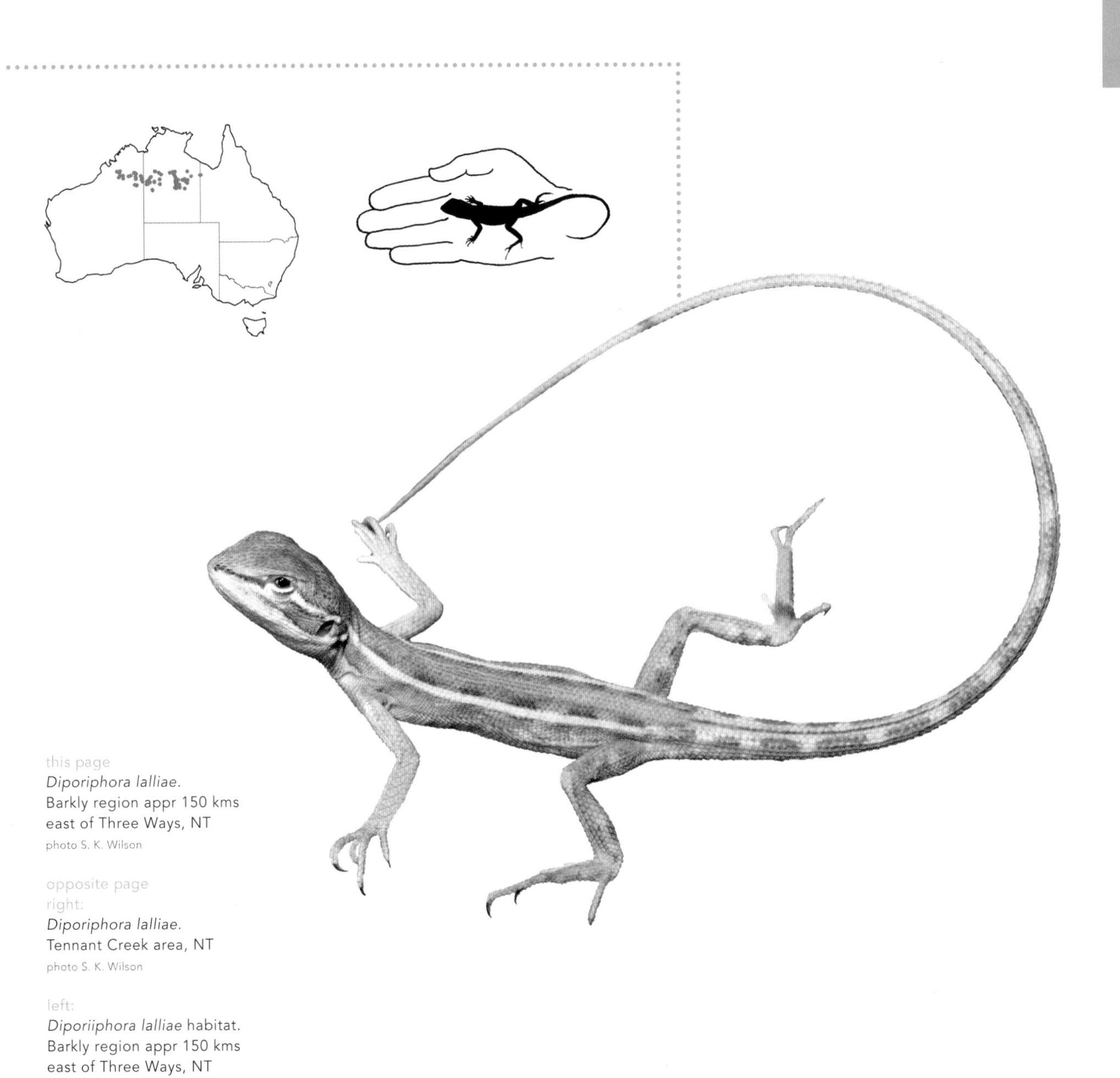

this page
*Diporiphora lalliae.*
Barkly region appr 150 kms
east of Three Ways, NT
photo S. K. Wilson

opposite page
right:
*Diporiphora lalliae.*
Tennant Creek area, NT
photo S. K. Wilson

left:
*Diporiiphora lalliae* habitat.
Barkly region appr 150 kms
east of Three Ways, NT
photo S. K. Wilson

## LINGA DRAGON

*Diporiphora linga* Houston, 1977

**DESCRIPTION:** SVL 55 mm. Robust with stout limbs and long tail. Gular and scapular folds present, postauricular fold absent. No crests and homogeneous dorsal scales with keels parallel to midline. Dorsal colour reddish-brown with 6–9 dark bands, no vertebral stripe, and usually some indication of pale, often yellowish, dorsolateral stripes, which can be very prominent in some individuals, particularly adult males. Males with breeding colour may have a prominent red flush on base of tail. Sides of body reddish-brown with pale flecks. Ventral surface pale and usually plain. Femoral pores 0; preanal pores 2.

**KEY CHARACTERS:** Similar to *D. reginae* but has a strong (versus weak) scapular fold and dark bars on the dorsal surface.

**DISTRIBUTION AND ECOLOGY:** *Acacia* and spinifex deserts of the southern interior of WA, extending into SA, in the southern portions of the Great Victoria Desert.

**BIOLOGY:** A semi-arboreal species that often perches atop *Triodia* hummocks. Little known about biology.

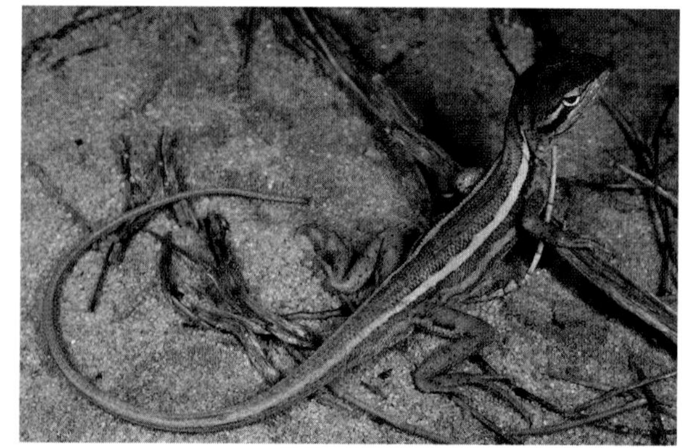

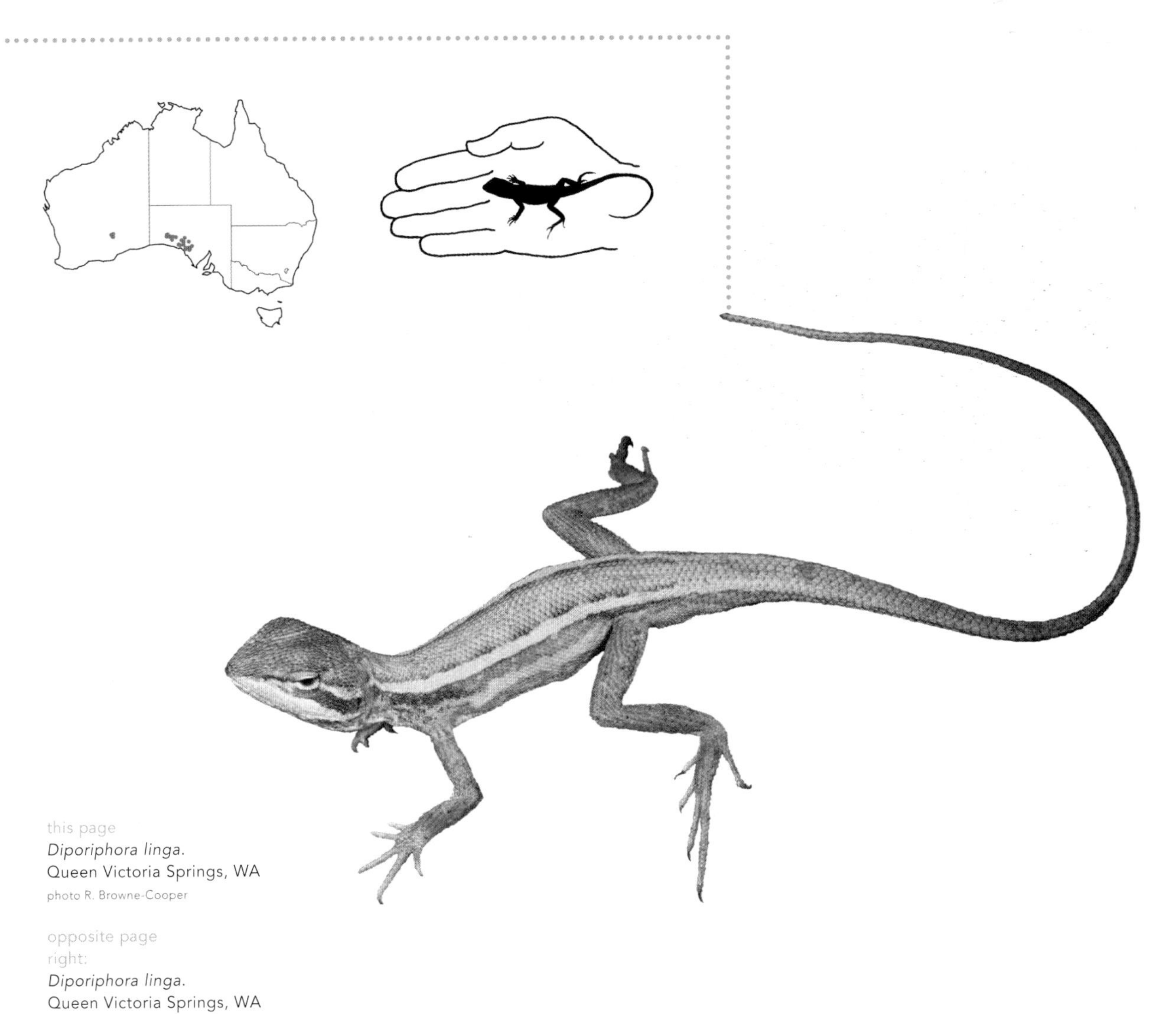

this page
*Diporiphora linga*.
Queen Victoria Springs, WA
photo R. Browne-Cooper

opposite page
right:
*Diporiphora linga*.
Queen Victoria Springs, WA
photo B. Maryan

left:
*Diporiphora linga* habitat.
Queen Victoria Springs area, WA
photo G. Gaikhorst

*Diporiphora linga.*
Queen Victoria Springs, WA
photo R. Browne-Cooper

*Diporiphora linga.*
Wirrula, SA
photo A. Elliott

## YELLOW-SIDED TWO-LINED DRAGON

*Diporiphora magna* Storr, 1974

DESCRIPTION: SVL 68 mm. Moderately gracile with long limbs and tail. Single canine on either side of upper jaw. Gular fold absent. Strong postauricular and scapular folds. Homogeneous, strongly keeled dorsal and ventral scales. Granular scales in armpit, extending over arm but not extending onto sides of neck. Pattern variable, from strongly patterned to plain. Patterned individuals have dark brown bands, often offset on either side of a broad greyish vertebral stripe, intersected by pale dorsolateral stripes. Plain individuals retain the pale dorsolateral stripes. Granular scales in armpit are dark brown, flanks cream to grey or light brown with little patterning. No white markings on face, labial scales speckled with light brown flecks. Ventral surface cream, usually plain but some individuals have faint longitudinal stripes on throat. Breeding males tend to lose some dorsal patterning, having a yellow "wash" over head and upper body with a large back patch in armpit extending onto shoulder. Femoral pores 0; preanal pores 4.

KEY CHARACTERS: Overlaps numerous other *Diporiphora* species across northern Australia. Differs from *D. gracilis* and *D. margaretae* in having strong postauricular and scapular folds (versus weak or absent), and lacking scattered white scales on a dark background on flanks. Differs from *D. pindan* in lacking white stripe between eye and ear. Differs from *D. bilineata* in having a postauricular fold (versus absent), having homogeneous dorsal scales (versus rows of enlarged vertebral scales) and lacking dark flanks with scattered white scales. Differs from *D. granulifera* in having granular scales restricted to scapular fold (versus granular scales extending over shoulder, along scapular fold and anteriorly onto neck). Differs from *D. lalliae* in lacking gular fold (versus present) and possessing granular scales in armpit. Differs from *D. albilabris, D. sobria, D. perplexa* and *D. bennettii* in lacking a gular fold and having single canine on either side of the upper jaw and further from *D. albilaris* and *D. sobria* in having homogeneous dorsal scales (versus dorsolateral rows enlarged on *D. albilabris* and raised on *D. sobria*).

DISTRIBUTION AND ECOLOGY: A habitat generalist occurring widely across the tropical savannah woodlands and grasslands of northern Australia, from east of Derby, WA, through NT to western Qld. Does not extend down into the arid zone, and in the far north, it is replaced by *D. margaretae* in the Kimberley and *D. bilineata* in NT, and in the east by *D. granulifera* in Qld.

BIOLOGY: Little known about biology. Seen to perch on low vegetation, rocks or termite mounds.

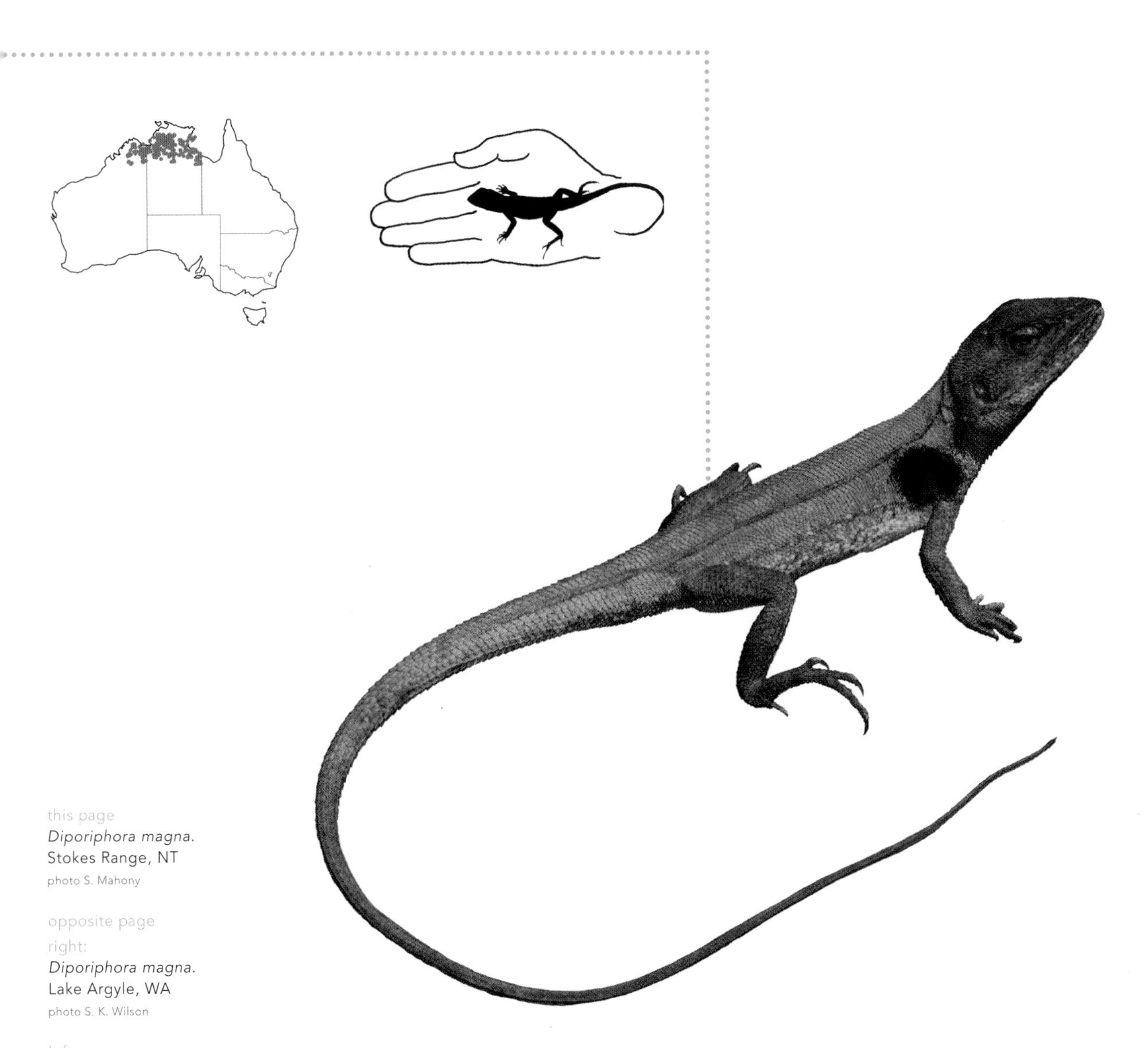

this page
*Diporiphora magna*.
Stokes Range, NT
photo S. Mahony

opposite page
right:
*Diporiphora magna*.
Lake Argyle, WA
photo S. K. Wilson

left:
*Diporiphora magna* habitat.
Mabel Downs Station, WA
photo S. K. Wilson

this page
*Diporiphora magna.*
Top Springs, NT
photo R. Glor

opposite page:
*Diporiphora magna.*
Theda Station, WA
photo B. Schembri

## MARGARET'S TWO-LINED DRAGON

*Diporiphora margaretae* Storr, 1974

**DESCRIPTION:** SVL 48 mm. Moderately gracile with long limbs and tail. Single canine on either side of upper jaw. Gular fold absent. Postauricular and scapular folds weak or absent. Homogeneous strongly keeled dorsal scales. Longitudinal series of raised but not enlarged pale paravertebral and dorsolateral scales at shoulder. Granular scales in armpit, extending over arm onto neck to posterior edge of scapular fold. Dorsal colour pattern variable. Patterned individuals, cream, grey, light to dark brown with broad greyish vertebral stripe, white dorsolateral stripes and 5–6 irregular dark brown bands reaching dorsolateral stripes but not usually extending beyond them. Granular scales on flanks around arm usually dark brown, extending posteriorly onto flanks. Flanks speckled with scattered white scales on a dark background. No white markings on face, labial scales speckled with dark brown flecks. Ventral surface cream sometimes with a few scattered flecks of light brown on belly, throat and head. Femoral pores 0; preanal pores 4.

**KEY CHARACTERS:** Potentially overlaps a number of other *Diporiphora* species in the far north Kimberley. Differs from *D. magna* in having weak or absent postauricular and scapular folds (versus strong folds) and having flanks speckled with scattered white scales on a dark background. Differs from *D. albilabris*, *D. sobria*, *D. perplexa* and *D. bennettii* in lacking a gular fold and having single canine on either side of the upper jaw and further from *D. albilaris* and *D. sobria* in having homogeneous dorsal scales (versus dorsolateral rows enlarged on *D. albilabris* and raised on *D. sobria*).

**DISTRIBUTION AND ECOLOGY:** Restricted to the far north Kimberley region, WA. A habitat generalist occurring in tropical savannah woodlands and grasslands. Also occupies coastal vegetation and has been found on the beach at Kulumbaru on the northern Kimberley coast.

**BIOLOGY:** Little known about biology. Seen to perch on low vegetation, rocks or termite mounds.

**NOTES:** *Diporiphora margaretae* was originally described as a subspecies of *D. bilineata* (Storr, 1974). However, genetic work has shown that this lineage is not closely related to *D. bilineata* and recent taxonomic work has raised *D. margaretae* to species status (Melville *et al*, 2019).

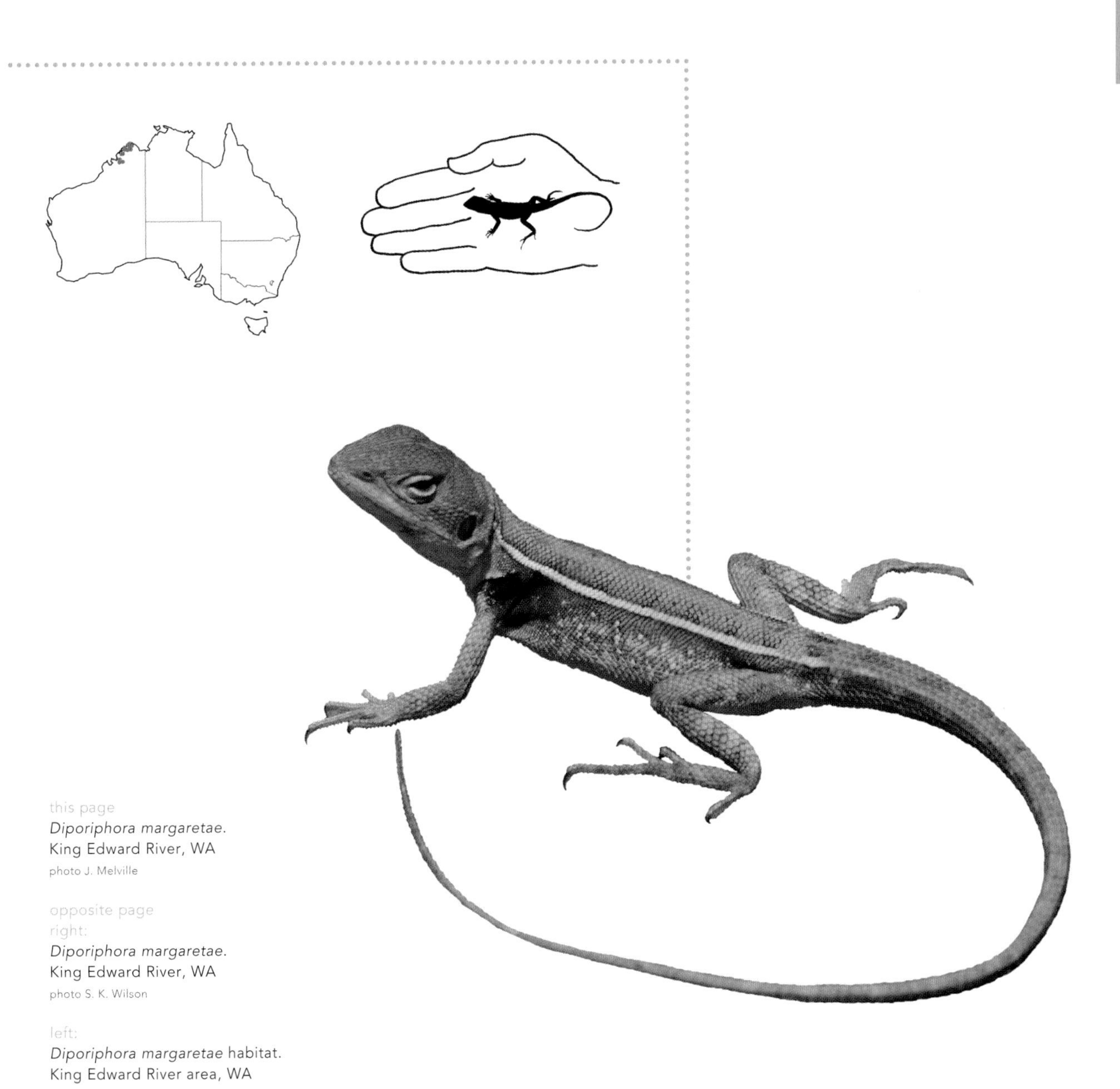

this page
*Diporiphora margaretae*.
King Edward River, WA
photo J. Melville

opposite page
right:
*Diporiphora margaretae*.
King Edward River, WA
photo S. K. Wilson

left:
*Diporiphora margaretae* habitat.
King Edward River area, WA
photo S. K. Wilson

this page
*Diporiphora margaretae*.
King Edward River, WA
photo J. Melville

opposite page:
*Diporiphora margaretae*.
Mt Elizabeth Station, WA
photo J. Vos

## NOBBI DRAGON

*Diporiphora nobbi* Witten, 1972

**DESCRIPTION:** SVL 72 mm. Robust with moderately long limbs and tail. Single canines on either side of upper jaw. Gular fold present. Strong postauricular and scapular folds. Dorsal scales heterogeneous with five rows of enlarged spines on back and enlarged spines around the postauricular fold. Scales on thighs strongly heterogeneous with spinose scales along the dorsal surface and small scales towards the posterior edges. Patterning variable. Dorsal colouration light brown to slate grey. A series of about 6 dark brown diamonds extending laterally between broad unpatterned pale grey or a dark charcoal vertebral stripe and prominent pale grey, cream to bright yellow dorsolateral stripes. Flanks vary in colour from chestnut brown to dark charcoal grey. White stripe between eye and ear, often with wide brown stripe below. Ventral surface cream with some grey flecks and lines particularly towards the flanks. Breeding males develop pink flush on base of tail, yellow wash over lower flanks, and dorsolateral strips are bright yellow. Femoral pores 2–4 on each side; preanal pores 3–4 on each side.

**KEY CHARACTERS:** Distinguished from *D. phaeospinosa* in having fewer femoral and preanal pores (2–4 femoral and 3–4 preanal on each side versus 4–6 femoral and 4–6 preanal), fewer nuchal spines and breeding males lacking a black throat. Overlaps extensively with *D. australis* and may overlap with *D. jugularis* to the north. Differs in being larger, having prominent spinose scales on thigh and neck, five rows of enlarged spines on back, and having more preanal and femoral pores. Differs further from *D. jugularis* in having a gular fold.

**DISTRIBUTION AND ECOLOGY:** Widespread along the east-coast of Australia, adjacent inland regions, west of Great Dividing Range and extending into north-western Vic. and eastern SA. A generalist species that occurs in a wide range dry, semi-arid and arid environments, including open forests, woodlands and shrublands. In Vic and SA it occurs in mallee woodlands. Ecological research in NSW mallee woodlands found that *D. nobbi* is sensitive to habitat fragmentation, as it is not found in grazed areas, roadsides or linear remnants (Driscoll, 2004). They have also been found to be sensitive to logging of eucalypt forests in NSW, being disadvantaged by increases in understorey and mid-canopy vegetation after logging (Kavanagh & Stanton, 2005). In Vic, *D. nobbi* has been found to decline in the short-term following fire, instead favouring vegetation about 15 years post-fire (Robertson *et al*, 2005).

**BIOLOGY:** A territorial species, using low fallen timber, rocks and dirt mounds as perches. Females usually lays 2 clutches each breeding season (Oct-Feb). The sex of offspring is genetically determined (Ezaz *et al*, 2009).

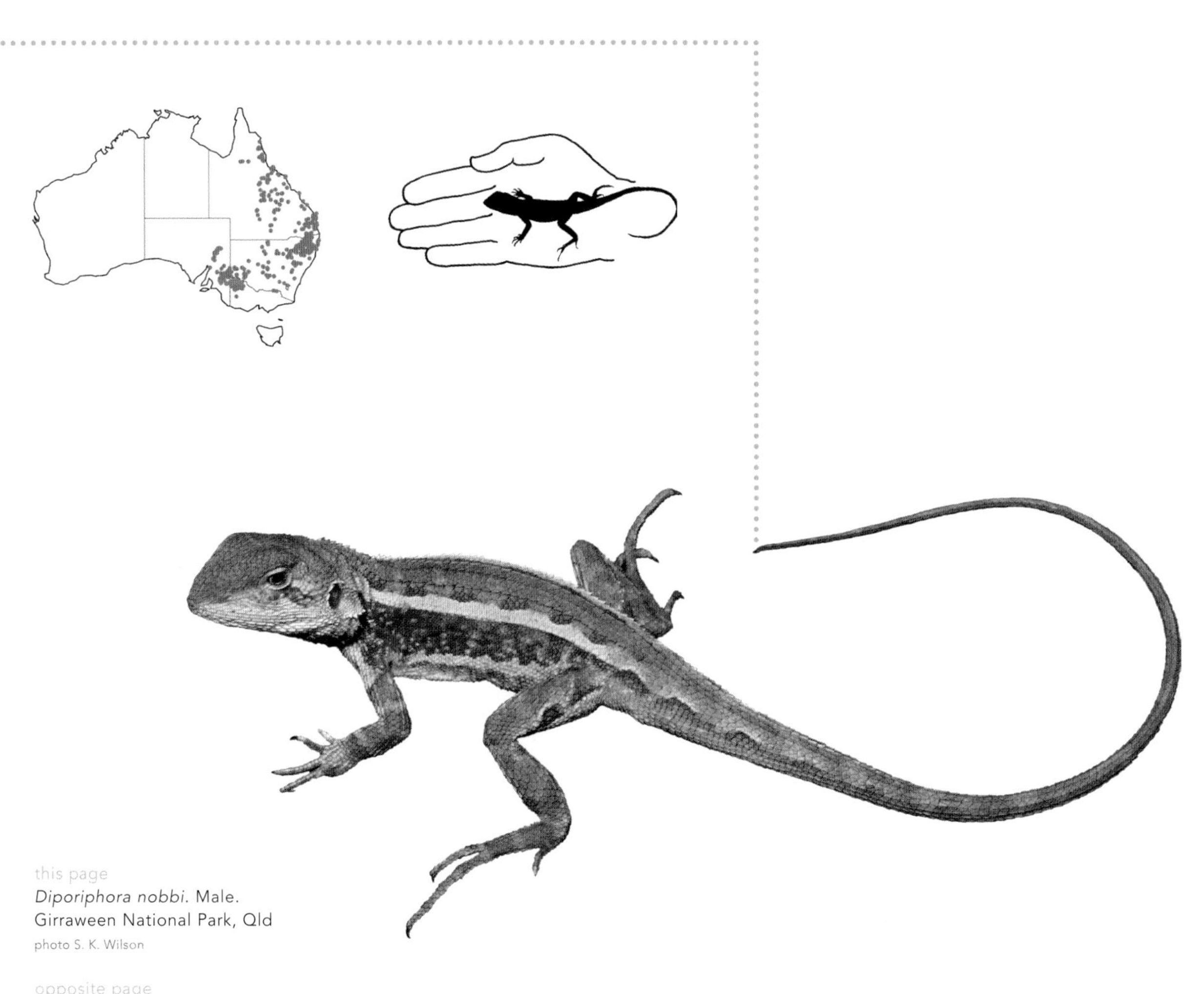

this page
*Diporiphora nobbi*. Male.
Girraween National Park, Qld
photo S. K. Wilson

opposite page
right:
*Diporiphora nobbi*. Male.
Wyperfeld National Park, Vic
photo N. Clemann

left:
*Diporiphora nobbi* habitat.
Pilliga State Forest, NSW
photo S. K. Wilson

*Diporiphora nobbi*. Male.
Aramac area, Qld
photo S. K. Wilson

*Diporiphora nobbi*. Male.
Girraween National Park, Qld
photo S. K. Wilson

## PALE TWO-PORED DRAGON

*Diporiphora pallida* Melville, Smith Date, Horner & Doughty, 2019

**DESCRIPTION:** SVL 46 mm. Small with robust head, prominent ridges from above eyes to snout and relatively short tail and limbs. Gular and scapular folds present; postauricular fold absent. Single canine on each side of upper jaw. Strongly keeled, homogeneous dorsal and ventral scales, with dorsal keels parallel to the midline. Scales in armpit small but not granular. Pale cream and light brown with little patterning; broad pale grey vertebral stripe, approximately 3 scales wide and broad yellow-cream dorsolateral stripes, approximately 3 scales wide, running from back of head to base of tail but interrupted by background colour. No dark spot in armpit or on neck, and no white markings on face. Faint pink flush on tail and hind limbs. Ventral surface white and unpatterned. Femoral pores 0; preanal pores 2.

**KEY CHARACTERS:** *D. pallida* is morphologically distinctive and readily distinguished from other *Diporiphora* species in the north-west Kimberley region by prominent ridges above eyes.

**DISTRIBUTION AND ECOLOGY:** A single animal presumed to be an adult, recorded from the Mitchell Plateau in the north-west Kimberley region, WA. It was found perched in spinifex grass on a rocky outcrop. Little is known of this species but it appears to be associated with spinifex grasses on rocky substrates.

**BIOLOGY:** Little known about biology.

**COMMENTS:** Morphological distinctiveness and genetic analyses indicate that *D. pallida* is not related to other *Diporiphora* in the Kimberley.

*Diporiphora pallida.*
Mitchell Plateau, WA
photo J. Melville

opposite page:
*Diporiphora pallida* habitat.
Mitchell Plateau, WA
photo R. Koch

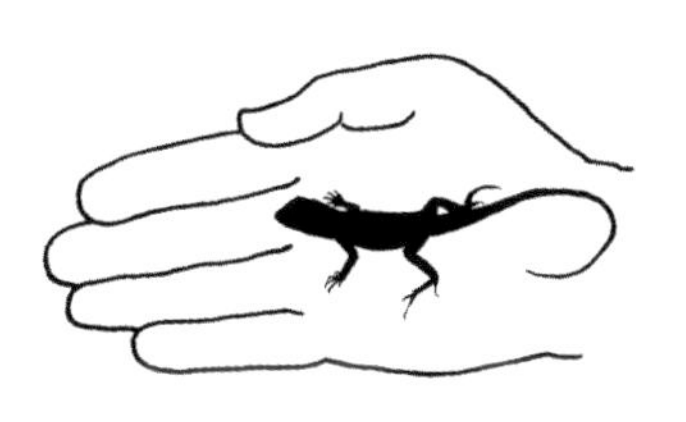

### GREY-STRIPED WESTERN DESERT DRAGON

*Diporiphora paraconvergens* Doughty, Kealley & Melville, 2012

**DESCRIPTION:** SVL 52 mm. Slender and elongate with weak gular and postauricular folds, and scapular fold present. No crests and homogeneous dorsal scales with keels on upper dorsal surface converging to midline. Dorsal colour dull light red to pink with very faint vertebral stripe. Grey dorsolateral stripes start on neck and continue to tail and narrow pale grey lateral stripe commences in front of arm and continues to leg and weakly onto side of tail. Nuchal region has a yellow wash and a pale stripe from eye to front of tympanum; labials and eyelids pale. Ventral surface pale with a pair of wide brown-grey stripes from snout, separating through gular region, converging on neck, widening and separating on body before converging near hindlimbs, then forming a grey background colour on underside of tail. Also two dark stripes under jaw, extending to sides of neck. Femoral pores 0; preanal pores 2.

**KEY CHARACTERS:** Differs from *D. vescus*, *D. pindan*, *D. valens* and *D. adductus* in having keels on dorsal surface converging towards midline. Differs further from *D. pindan* in having gular and postauricular folds (versus absent). Differs from *D. winneckei* in having two preanal pores (versus absent), no reduction of scale size on upper lateral region (versus upper lateral scales conspicuously reduced in size relative to surrounding scales), a longer tail, and having thin (versus thick) dark markings either side of midline on gular region.

**DISTRIBUTION AND ECOLOGY:** Occurs mainly on sand dunes, usually seen perching on vegetation, including *Acacia*, *Grevillea*, and *Triodia*. Extends across north-western sand deserts, including Great Sandy Desert and has been found in dunefields of the northern Pilbara. Also occurs around Lake Mackay near the WA-NT border and along the north-western portion of the Great Victoria Desert. Extends into north-western SA, and probably south-western NT.

**BIOLOGY:** A semi-arboreal species for which little known about biology.

opposite page left:
*Diporiphora paraconvergens* habitat.
Maroochydore mine area, WA
photo G. Harold

opposite page right:
*Diporiphora paraconvergens*.
Maroochydore mine area, WA
photo G. Harold

below:
*Diporiphora paraconvergens*.
Lake Mackay, WA
photo G. Gaikhorst

## KIMBERLEY TWO-PORED ROCK DRAGON

*Diporiphora perplexa* Melville, Smith Date, Horner and Doughty, 2019

**DESCRIPTION:** SVL 65 mm. Moderately robust with long tail and limbs. Two enlarged canines on each side of upper jaw, with posterior canine extremely enlarged. Dark pigment "smear" on posterior of tympanum spreading onto neighbouring head scales. Gular, postauricular and scapular folds present, although relatively weak. Scales in armpit small but not granular, extending over arm and onto neck, past scapular fold. Dorsal body scales homogeneous and strongly keeled. Variable patterning from strongly patterned to unpatterned. Pale dorsolateral stripes usually present. Patterned individuals also have 5–6 dark bands intersected by a narrow white, cream or grey vertebral stripe and a black patch enclosing pale flecks on shoulder and in armpit. Individuals with little patterning, usually adult males with breeding colours, are bright yellow with pink flush on tail and hindlimbs, and a large black patch in armpit, extending up onto shoulder, then fading posteriorly to dark speckled appearance. No distinct patterning on head, upper labials flecked with light brown and cream, with no pale labial stripe. Ventral surface plain cream to white, with some individuals having a few scattered brown scales on throat. Femoral pores 0 (although a few individuals have 1); preanal pores 2–4.

**KEY CHARACTERS:** Differs to *D. bennettii* in having a long tail and limbs and dorsolateral stripes (versus very short tail and limbs). Differs from *D. albilabris* and *D. sobria* in mostly lacking femoral pores, having no white or pale stripes on head (versus present on upper lip or between eye and ear), no stripes under chin and homogeneous dorsal scales (versus dorsolateral rows enlarged on *D. albilabris* and raised on *D. sobria*). Differs from *D. margaretae*, *D. pindan* and *D. magna* in having a gular fold, a black spot on tympanum and two canine teeth on each side of upper jaw (versus one on each side).

**DISTRIBUTION AND ECOLOGY:** Occurs on large rock outcrops and escarpments in north-western Australia. Individuals have been found in the northern Kimberley Plateau, south to Fitzroy Crossing and to western NT in the Jasper Gorge area. This is a true rock-dwelling specialist that has only been found associated with large outcrops.

**BIOLOGY:** Little known about biology. Has only been observed perching on rock outcrops and rocky escarpments.

**NOTES:** This species has previously been confused with *D. bennettii* and has usually been depicted in field guides as *D. bennettii* but both genetics and examination of the original specimens demonstrate clear differences (Melville *et al*, 2019).

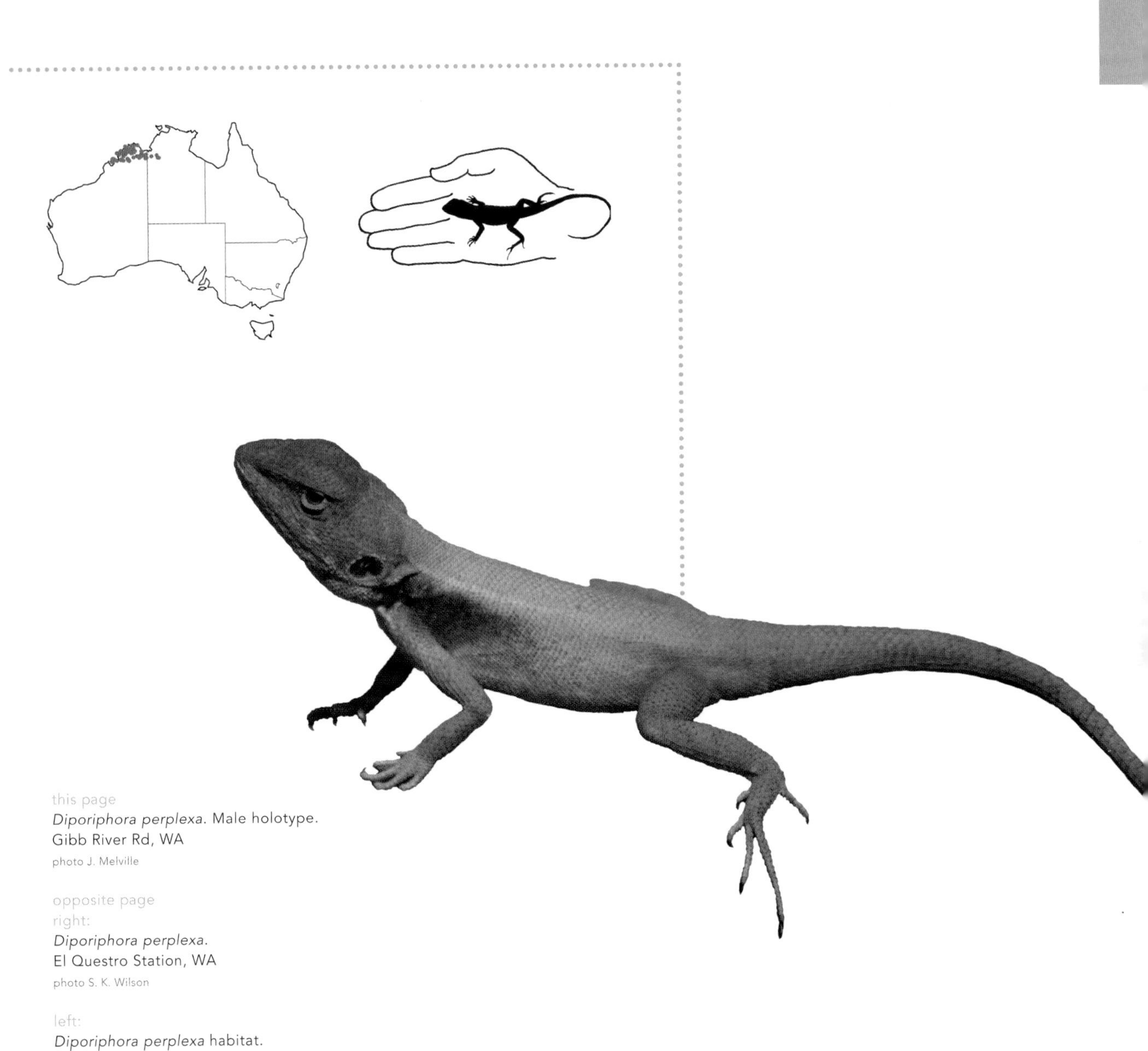

this page
*Diporiphora perplexa*. Male holotype.
Gibb River Rd, WA
photo J. Melville

opposite page
right:
*Diporiphora perplexa*.
El Questro Station, WA
photo S. K. Wilson

left:
*Diporiphora perplexa* habitat.
Jasper Gorge area, NT
photo R. Koch

*Diporiphora perplexa*. Breeding male.
Mt Elizabeth Station, WA
photo J. Vos.

opposite page:
*Diporiphora perplexa*. Female paratype.
King Edward River, WA
photo J. Melville.

## BLACK-SPINED DRAGON

*Diporiphora phaeospinosa* Edwards & Melville, 2011

**DESCRIPTION:** SVL 66 mm. Robust with moderately long limbs and tail. Single canine on either side of upper jaw. Gular fold present. Strong postauricular and scapular folds. Large nuchal spines (approx. 9) spanning back of head in an irregular row, separate to dorsal spines. Dorsal scales heterogeneous with five rows of enlarged spines on back and enlarged spines around scapular fold. Row of enlarged dorsolateral scales running along back, and enlarged spinose scales scattered randomly over dorsolateral area. Scales on thighs strongly heterogeneous with spinose scales along the dorsal surface and small scales towards posterior edge. Patterning variable. Light brown to slate grey above with a series of about 6 dark brown diamonds between broad unpatterned pale grey or dark charcoal vertebral stripe and prominent cream to yellow, sometimes deeply notched, dorsolateral stripes. Flanks range from chestnut brown to dark charcoal grey. White stripe between eye and ear, often with wide brown stripe below. Lip scales white, contrasting with surrounding background colour. Ventral surface cream with some grey flecks and lines particularly towards the flanks. Base of tail flushed with pink. Breeding males have a black throat extending from upper chest to chin. Femoral pores 4–6 on each side; preanal pores 4–6 on each side.

**KEY CHARACTERS:** Distinguished from the Nobbi Dragon (*D. nobbi*) in having more femoral and preanal pores on each side (4–6 femoral, and 4–6 preanal pores versus 2–4 femoral and 3–4 preanal pores), more nuchal spines, a row of enlarged dorsolateral scales running down the back and a black throat on breeding males. Differs from the Tommy Round-head (*D. australis*) in being larger, having prominent spinose scales on thigh, five rows of enlarged spines on back and more femoral and preanal pores.

**DISTRIBUTION AND ECOLOGY:** Currently known from four localities in central western Qld, representing two discrete regions, the Carnarvon Gorge/Bigge Range area, and the vicinity of Blackdown Tablelands. It is currently unknown whether the distribution of the species is continuous, or whether these discrete regions represent disjunct populations. It occupies dry sclerophyll forests, woodlands and heaths, often in association with sandstone outcrops.

**BIOLOGY:** Commonly seen perching on rocks and low timber. Little is known about the biology of this species but assumed to be similar to *D. nobbi*.

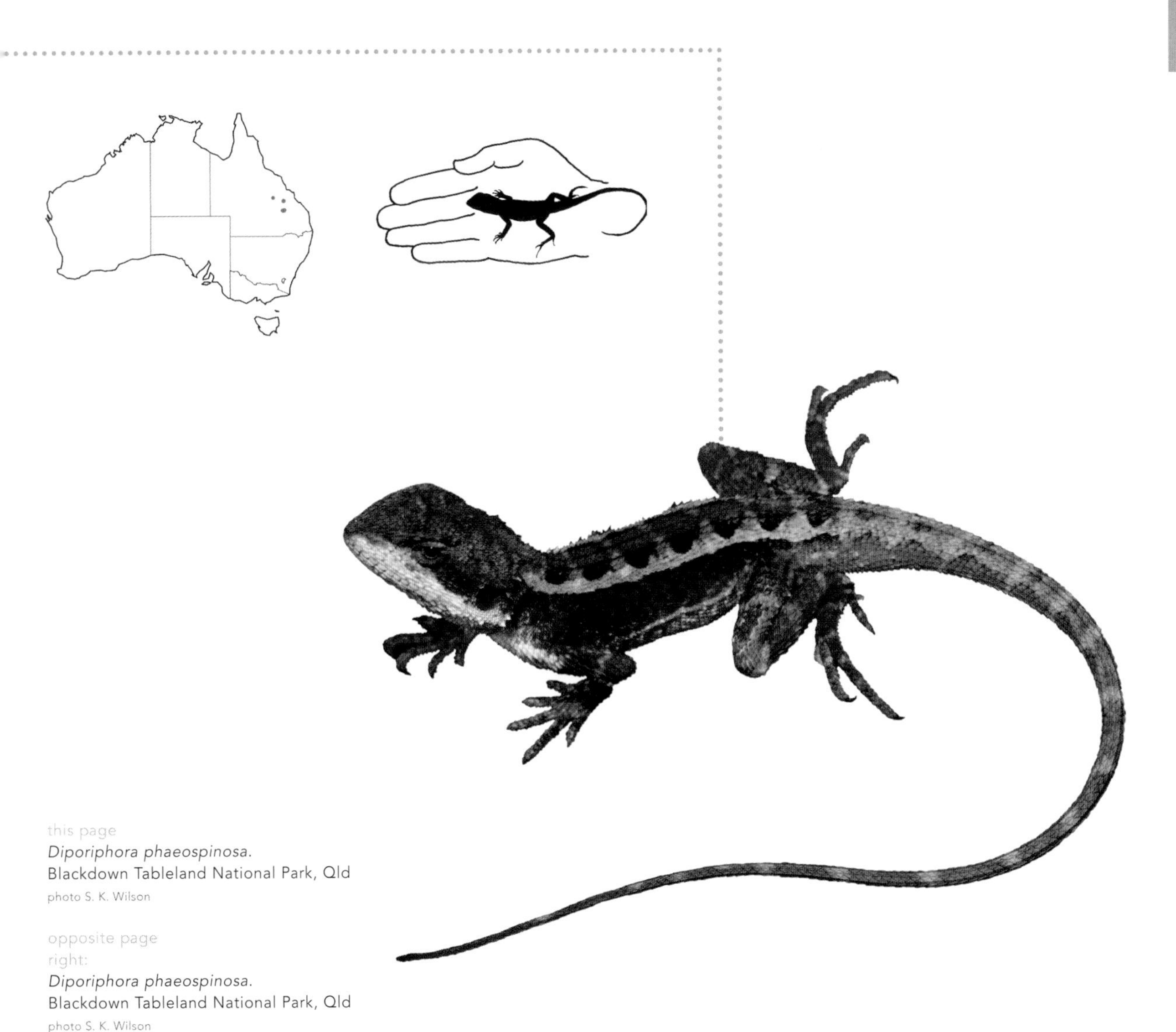

this page
*Diporiphora phaeospinosa.*
Blackdown Tableland National Park, Qld
photo S. K. Wilson

opposite page
right:
*Diporiphora phaeospinosa.*
Blackdown Tableland National Park, Qld
photo S. K. Wilson

left:
*Diporiphora phaeospinosa* habitat.
Blackdown Tableland National Park, Qld
photo S. K. Wilson

*Diporiphora phaeospinosa*. Male.
Blackdown Tableland National Park, Qld
photo S. K. Wilson

opposite page:
*Diporiphora phaeospinosa*. Male.
Blackdown Tableland National Park, Qld
photo S. K. Wilson

## PINDAN DRAGON

*Diporiphora pindan* Storr, 1979

DESCRIPTION: SVL 60 mm. Slender with long limbs and tail, A single canine on each side of the upper jaw. No gular or postauricular folds, and a weak scapular fold. Homogeneous dorsal scales with keels parallel to midline or at most only weakly converging towards midline and no crests. Dorsal colour and pattern vary from plain to complex, but most individuals have a prominent pale stripe, often with dark brown border, between eyes and ears. This stripe may be absent. Animals with complex patterning are light brown with pale grey vertebral stripe 2–3 scales wide and pale yellow-white dorsolateral stripes. Flanks have mosaic of light and dark scales that transition to pale ventral colour, but usually no well-defined lateral stripe. Animals with simple patterns are uniform brown to black with highly contrasting pale yellow dorsolateral stripes (usually males) or lighter uniform colour with no dorsolateral stripes (usually females). Males often have a large black circular mark on the sides, posterior to the arms. Ventral surface white without patterning or with two dark longitudinal stripes from tip of snout down length of body, sometimes merging to form a single dark ventral patch. Femoral pores 0; preanal pores 0–4.

KEY CHARACTERS: In the northern Pilbara coastal region *D. pindan* overlaps with *D. vescus* and *D. paraconvergens*, from which it differs in lacking gular and postauricular folds. In the Kimberley *D. pindan* overlaps with *D. gracilis* and the far western range of *D. magna* and *D. lalliae*. Differs from *D. gracilis* in having weakly keeled dorsal scales (versus strongly keeled) and usually having a well-defined white stripe between eye and ear (versus absent). It differs from *D. lalliae*, in lacking gular and postauricular folds, and from *D. magna* in lacking a postauricular fold, lacking granular scales in armpit and having prominent pale stripe between eye and ear. It may overlap with the far western distribution of *D. sobria* but can be distinguished by lacking gular and postauricular folds, having single canine (versus two) on each side of upper jaw and having homogeneous dorsal scales (versus raised dorsolateral rows).

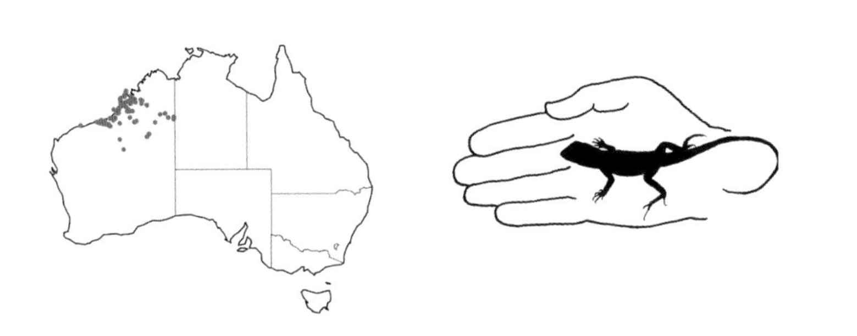

opposite page right:
*Diporiphora pindan.*
Point Smith area, WA
photo R. Glor

opposite page left:
*Diporiphora pindan* habitat.
Point Smith area, WA
photo R. Glor

below:
*Diporiphora pindan.*
Cape Leveque, WA
photo S. K. Wilson

**DISTRIBUTION AND ECOLOGY:** Occurs on sandy soils vegetated with *Triodia*, grasses, and *Acacia*, extending from the Dampier Peninsula in the south-west Kimberley, south through the Great Sandy Desert to the Pilbara coast as far west as Karratha and east to the Tanami Desert.

**BIOLOGY:** Perches on vegetation in the day and while asleep at night; also observed perching on low termite mounds and on small rocks. Individuals have been found sheltering under low ground cover and have even been dug from burrows. Otherwise little is known about its biology.

### PLAIN-BACKED TWO-LINED DRAGON

*Diporiphora reginae* Glauert, 1959

**DESCRIPTION:** SVL 55 mm. Robust with stout limbs. Gular fold present, scapular fold weak to absent and postauricular fold absent. Dorsal scales homogeneous with keels parallel to midline and no crests. Dorsal colour reddish-brown with no bands or vertebral stripe but usually some indication of yellow dorsolateral stripes. These can be very prominent in some individuals, particularly adult males. Sides of body reddish-brown with pale flecks. Ventral surface white and usually plain. Breeding males may have a prominent red flush on the base of the tail. Femoral pores 0; preanal pores 2.

**KEY CHARACTERS:** Similar to *D. linga* but has a weak scapular fold and lacks any dark dorsal bands.

**DISTRIBUTION AND ECOLOGY:** Sandy deserts dominated by *Triodia* across the southern interior of WA, primarily in the Great Victoria Desert, but extending east to Moonaree Station, SA.

**BIOLOGY:** A semi-arboreal species that commonly perches atop *Triodia* hummocks. Otherwise little known about biology.

opposite page left:
*Diporiphora reginae* habitat.
Queen Victoria Springs area, WA
photo G. Gaikhorst

opposite page right:
*Diporiphora reginae*.
Great Victoria Desert, WA
photo R. Lloyd

below:
*Diporiphora reginae*.
Moonaree Station SA
photo A. Elliott

## NORTHERN TWO-PORED ROCK DRAGON

*Diporiphora sobria* Storr, 1974

**DESCRIPTION:** SVL 52 mm. Moderately robust with long limbs and tail. Two canines (double) on either side of upper jaw. Gular fold present. Postauricular weak to absent and scapular fold weak or moderate. Dorsal scales variable but mostly relatively large and homogeneous, between two dorsolateral rows of pale raised scales. Between these dorsolateral rows, the scales are relatively large, homogeneous and strongly keeled; on the outer sides of these dorsolateral rows, the scales are relatively small and keeled. Pale dorsolateral scale rows often absent in NT animals. Ventral scales homogeneous and strongly keeled. Dorsal and lateral colour brown and grey, with pattern absent or diffuse. Labial scales pale. Throat, torso and tail white or cream without pattern. Breeding males have strongly contrasting charcoal black, white and chestnut or orange-red colouring on head and upper body, with tail also having a pink flush in some individuals. Femoral pores 2; preanal pores 4.

**KEY CHARACTERS:** Wide distribution, overlapping numerous other *Diporiphora* species. Differs in having the following combination of characters: two canines on each side of upper jaw; a gular fold; two femoral pores; and having white labial scales. In Kimberley region, distinguished from *D. albilabris* in having dorsal scales with two pale dorsolateral rows of raised but not enlarged scales; between these raised rows, the scales are relatively large, homogeneous and strongly keeled (versus strongly heterogeneous scales between dorsolateral scale rows).

**DISTRIBUTION AND ECOLOGY:** Widely distributed from the southern and central Kimberley region, across NT south to the Katherine area and into western Qld. A habitat generalist occurring in tropical savannah woodlands and grasslands.

**BIOLOGY:** Little known about biology. Seen to perch on low vegetation, rocks or termite mounds.

**NOTES:** Following a recent taxonomic revision (Melville *et al*, 2019), this species is expanded to incorporate what was known as *D. arnhemica* and much of the distribution of *D. albilabris*. However it remains a diverse species complex that will need further taxonomic consideration.

opposite page left:
*Diporiphora sobria* habitat.
Kakadu Hwy, NT
photo P. Horner

opposite page right:
*Diporiphora sobria*. Male.
Lichfield National Park, NT
photo B. Shembri

below:
*Diporiphora sobria*.
Adelaide River, NT
photo R. Lloyd

# DIPORIPHORA

## SUPERB DRAGON

*Diporiphora superba* Storr, 1974

**DESCRIPTION:** SVL 93 mm. Extremely long, slender body, narrow head, very long thin limbs and digits, and extremely long slender tail, sometimes exceeding 4 times the length of SVL. No gular, scapular or postauricular folds. Body scales homogeneous; no enlarged crests, spines or tubercles but ventral scales larger with stronger keels than dorsals. Green to bright green on back and sides, sometimes with a broad, brown to rich reddish brown dorsal stripe. Yellow below. Femoral pores 0; preanal pores 2.

**KEY CHARACTERS:** The extremely slender build and bright green colour distinguish this species from all Australian dragons.

**DISTRIBUTION AND ECOLOGY:** Arboreal, inhabiting slender stems and foliage in shrubs, bushes and vine thickets along the edges of sandstone gorges in north-western Kimberley region, WA. The extremely long appendages are not well suited for terrestrial locomotion, and the Superb Dragon is somewhat clumsy on the ground.

**BIOLOGY:** This is a relatively slow-moving species, relying on the camouflage its green colouration affords it among the foliage. It moves with such deliberation that it takes a keen eye to detect the subtle tell-tale movement of leaves that may betray its presence. In many respects it is unique among Australian dragons, including their bright green colouration and its almost exclusive utilisation of foliage. Captive breeding of this dragon has recorded a clutch size of 5 (Weigel, 1989). When first laid, the eggs were small and thinly shelled then over the first 3 weeks of incubation the eggs swelled to approximately double their size and developed a white, presumably calciferous, leathery surface. It was suggested that this may be a breeding strategy to allow larger clutch sizes by laying small 'undeveloped' eggs. All hatchlings emerged from the eggs a light coppery brown colour and when placed on green foliage, they rapidly turned a light green colour.

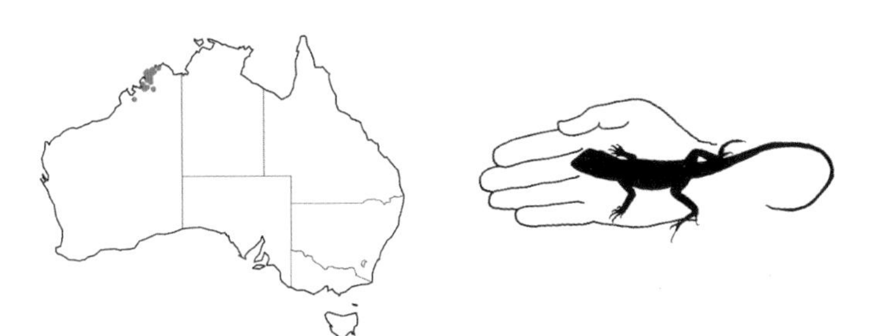

opposite page left:
*Diporiphora superba* habitat.
Surveyor's Pool, Mitchell Plateau, WA
photo S. K. Wilson

opposite page right:
*Diporiphora superba*.
Artesian Ranges, WA
photo P. Oliver

below:
*Diporiphora superba*.
Surveyor's Pool, Mitchell Plateau, WA
photo S. K. Wilson

*Diporiphora superba.*
Surveyor's Pool,Mitchell Plateau, WA
photo S. K. Wilson

opposite page:
*Diporiphora superba.*
Surveyor's Pool, Mitchell Plateau, WA
photo S. K. Wilson

## SOUTHERN PILBARA TREE DRAGON

*Diporiphora valens* Storr, 1979

**DESCRIPTION:** SVL 75 mm. Robust, with gular and scapular folds and a weak postauricular fold. Homogeneous dorsal scales with keels parallel to midline and no crests. Dorsal colour light to dark brown with very variable pattern, ranging from complex to plain. Complex patterned animals have a faint pale silver vertebral stripe, 9–11 dark brown bands between pale yellow dorsolateral stripes, which continue posteriorly onto tail or fade above pelvis, and narrow pale yellow lateral stripe usually discernible. Plain animals have greatly reduced pattern but retain dorsolateral stripes. Pale temporal stripe usually present. Ventral surface white and usually plain, although some more patterned individuals may have markings on medial gular region. Femoral pores 0; preanal pores 2–4.

**KEY CHARACTERS:** Differs from *D. vescus* in being more robust and having larger body size, spinier scalation, slightly shorter limbs and tail, darker background colour, and more dark bands. Differs from *D. paraconvergens* in having keels parallel to midline (versus converging towards midline).

**DISTRIBUTION AND ECOLOGY:** *Acacia* and *Triodia* dominated habitats in southern Pilbara region, WA. Mostly known from the Hamersley Range but there are isolated populations to the east and south-east in the Little Sandy Desert, possibly extending further into the interior through isolated rocky breakaway country.

**BIOLOGY:** A semi-arboreal species for which little known about biology.

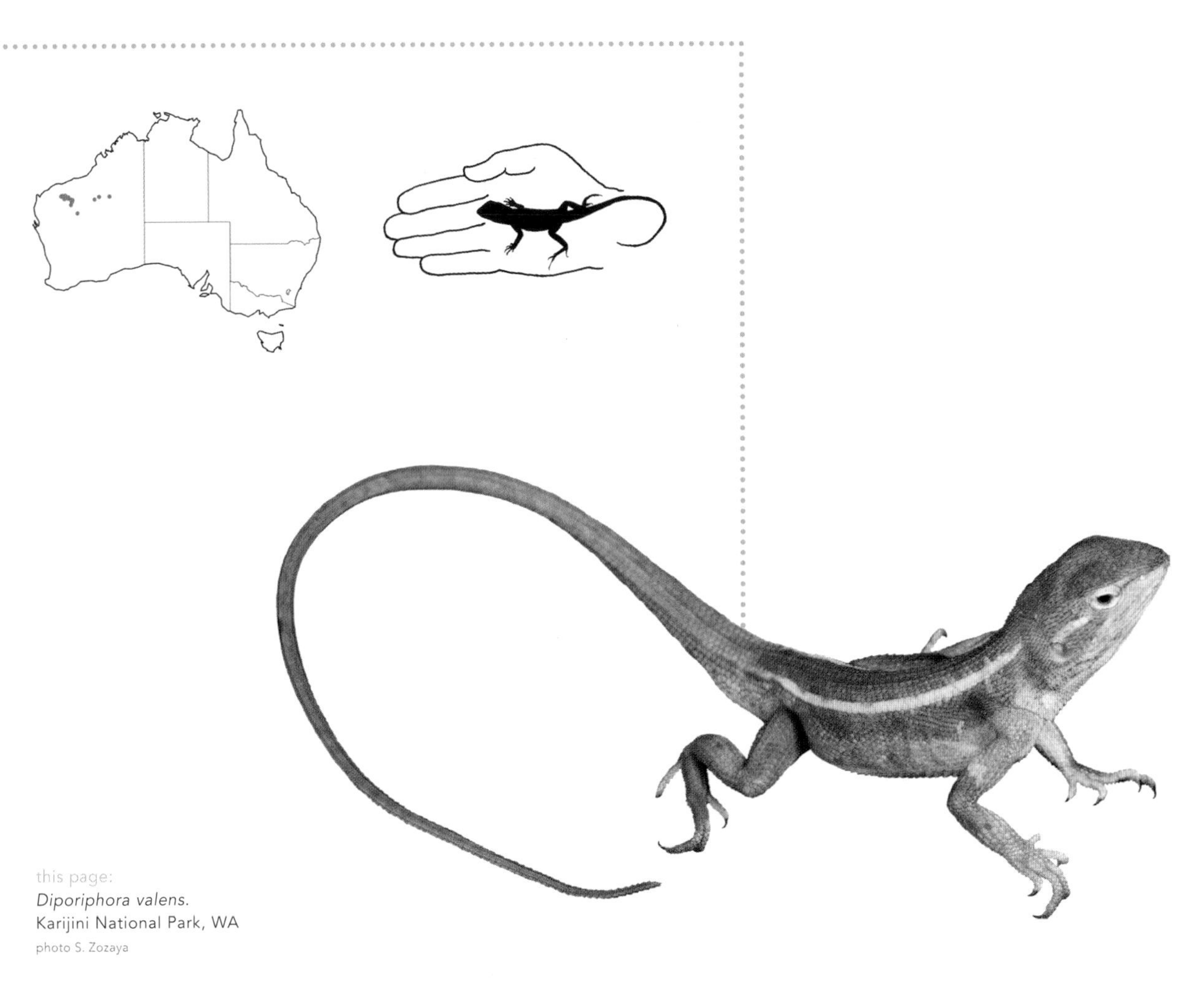

this page:
*Diporiphora valens.*
Karijini National Park, WA
photo S. Zozaya

opposite page
right:
*Diporiphora valens,*
Coondewanna Flats, WA
photo B. Maryan

left:
*Diporiphora valens* habitat.
Mt Robinson area, Hamersley Range, WA
photo S. Ford

*Diporiphora valens.*
Hamersley Range, WA
photo R. Lloyd

### NORTHERN PILBARA TREE DRAGON

*Diporiphora vescus* Doughty, Kealley & Melville, 2012

**DESCRIPTION:** SVL 60 mm. Moderately robust, with gular and scapular folds and a weak postauricular fold. Dorsal scales homogeneous with keels parallel to midline and no crests. Dorsal colour light brown with pattern ranging from complex to plain. Complex patterned animals have a faint pale silver vertebral stripe, about 6 dark brown bands between pale yellow dorsolateral stripes which continue posteriorly onto tail or fade above pelvis, and usually a narrow pale yellow lateral stripe. Plain patterned animals retain dorsolateral stripes. Pale temporal stripe is usually present. Breeding males develop a pinkish flush on hips and tail. Ventral surface white without patterning. Femoral pores 0; preanal pores 2–4.

**KEY CHARACTERS:** Differs from *D. valens* in having less robust build, smaller body size, less spiny scalation, slightly longer limbs and tail, lighter background colour, and fewer dark bands. The northern Pilbara coastal population overlaps with *D. pindan* and *D. paraconvergens*, from which it differs by having a gular and weak postauricular fold (absent on *D. pindan*) and dorsal scales with keels parallel to midline (converging towards midline on *D. paraconvergens*).

**DISTRIBUTION AND ECOLOGY:** Sandy and clayey alluvial soils and coastal sand dunes in the northern Pilbara region of WA, within the Chichester and the coastal Roebourne subregions.

**BIOLOGY:** A semi-arboreal species for which little known about biology. They are not commonly encountered but have been observed perching on fence posts.

**NOTES:** Recently listed as vulnerable under the international IUCN Red List (Doughty *et al*, 2017) due to pastoral activities and introduced grasses and predators.

opposite page left:
*Diporiphora vescus* habitat.
Port Hedland, WA
photo S. Zozaya

opposite page right:
*Diporiphora vescus*. Male.
Turner River, WA
photo R. Lloyd

below:
*Diporiphora vescus*. Female.
Port Hedland, WA
photo B. Schembri

*Diporiphora vescus*. Male.
Port Hedland, WA
photo S. Zozaya

## CANEGRASS DRAGON

*Diporiphora winneckei* Lucas & Frost, 1896

**DESCRIPTION:** SVL 60 mm. Elongate and slender with weak gular and postauricular folds, but strong scapular fold. A single canine on each side of the upper jaw. Dorsal scales homogeneous with keels converging slightly towards midline and no crests. Upper lateral scales conspicuously reduced in size relative to surrounding scales, and lower lateral scales enlarged with keels directed slightly dorsally. Dorsal colour light brown with pale grey-silver vertebral stripe, about 6 dark brown bands and prominent pale yellow-white dorsolateral stripes, which continue onto tail or fade near hindlimbs. Prominent pale stripe with dark brown border from behind eye, sometimes fading half-way to tympanum. Labials white, with pale stripe continuing to beginning of the neck. Ventral surface white with a medial brownish-grey to yellow stripe on gular region from near tip of snout, then separating into two stripes on body, converging again in front of hindlimbs, and continuing onto tail or fading. Some individuals lack the medial stripe in the gular region and instead have two yellow-grey stripes running parallel to jawline. Femoral pores 0; preanal pores 0.

**KEY CHARACTERS:** In western Qld, *D. winneckei* may overlap with *D. ameliae* from which it can be distinguished by the ventral stripes, having only two stripes that are well spaced at midbody and converging to meet in the pectoral and pelvic regions (versus four continuous or broken ventral stripes when present). Additionally, *D. winneckei* has a weak postauricular fold (versus present and spinose on *D. ameliae*). Differs from *D. paraconvergens* in having upper lateral scales conspicuously reduced in size relative to surrounding scales (versus no reduction of scale size on upper lateral region). Other species that potentially overlap *D. winneckei* include *D. lalliae* and *D. linga*, but these differ from *D. winneckei* in having strong gular folds and having preanal pores.

**DISTRIBUTION AND ECOLOGY:** Occurs in vegetation such as *Triodia*, and canegrass (*Zygochloa*) associated with sandy deserts, particularly sandridges. Distributed across eastern arid zone, including south-western Qld, north-eastern SA and southern NT.

**BIOLOGY:** A semi-arboreal species for which little known about biology. Have been observed perching on canegrass (*Zygochloa paradoxa*) on sandridges (Houston & Hutchinson, 1998), including on the very top branches of canegrass on the top of sandridges in the hottest part of the day, presumably to catch the wind for cooling. Also seen clinging to lower branches of canegrass at night. In spring and early summer an average active body temperature of 37.7°C has been recorded (Melville & Schulte, 2001).

opposite page right:
*Diporiphora winneckei*.
Windorah, Qld
photo S. K. Wilson

opposite page left:
*Diporiphora winneckei* habitat.
Ethabuka, Qld
photo S. K. Wilson

below:
*Diporiphora winneckei*. Gravid female.
Windorah, Qld
photo S. K. Wilson

*Gowidon longirostris.*
Barkly Region, NT.
photo S. K. Wilson

# GOWIDON

## LONG-NOSED DRAGON; LONG-NOSED WATER DRAGON

Genus *Gowidon* Wells & Wellington, 1984

*Gowidon longirostris* Boulenger, 1883

# GOWIDON

## LONG-NOSED DRAGON; LONG-NOSED WATER DRAGON

*Genus Gowidon* Wells & Wellington, 1984; *Gowidon longirostris* Boulenger, 1883

**DESCRIPTION:** Sole member of genus. SVL 145 mm. Moderately large with long slender limbs, very long tail, narrow dorsoventrally compressed head, long snout and distinct erectable nuchal crest comprising short thick spines. Dorsal scales homogeneous, with keels converging posteriorly toward midline. Pale to dark greyish brown with prominent pale dorsolateral stripes, pale stripe along lower jaw and reddish brown flush on face. One to three small white spots on a black background positioned directly behind the ear. About three reddish blotches along each side of ventral surface. Preanal pores 4–7; femoral pores 11–22.

**KEY CHARACTERS:** Differs from *Amphibolurus centralis* in having homogeneous dorsal scales (versus heterogeneous, with several rows of enlarged scales). Differs further, and from *Lophognathus*, by having more than 10 femoral pores (versus fewer than 10 femoral pores), one or more white spots on black background behind ear, relatively long snout, dorsoventrally compressed head, and a very long whip-like tail.

**DISTRIBUTION AND ECOLOGY:** Arid western interior of Australia. Semi-arboreal, occurring in a broad range of habitats. Usually associated with tree-lined inland ephemeral watercourses, gorges and river beds. Also occurs in sandy desert areas, where it is associated with trees such as snappy gums, mallees and desert oaks.

**BIOLOGY:** Often seen perching on fallen tree debris, rocks and tree branches. When disturbed this lizard will run bipedally, with body held upright, ascending the nearest tree. Males are territorial and can be seen displaying with head bobbing, hand waves and tail flicking, while raising their erectable nuchal crests.

**NOTES:** Genetic work has confirmed that *Gowidon longirostris* is unrelated to *Amphibolurus* and *Lophognathus* species (Melville *et al*, 2011).

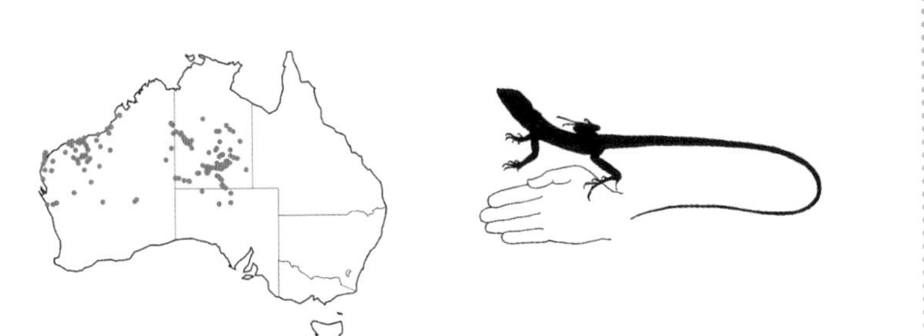

opposite page left:
*Gowidon longirostris* habitat.
Trephina Gorge, NT
photo S. K. Wilson

opposite page right:
*Gowidon longirostris.*
Ormiston Gorge, NT
photo R. Glor

below:
*Gowidon longirostris.*
Barkly Region, NT
photo S. K. Wilson

*Intellagama lesueurii howittii.*
Genoa Falls, Vic
photo S. K. Wilson

# INTELLAGAMA

## WATER DRAGON

*Intellagama* Wells and Wellington, 1985 *Intellagama lesueurii* Gray, 1831

# INTELLAGAMA

## WATER DRAGON

*Intellagama* Wells and Wellington, 1985, *Intellagama lesueurii* Gray, 1831

**DESCRIPTION:** Sole member of genus. A very large dragon (up to SVL 245 mm), with prominent spiny nuchal and vertebral crests, long powerful limbs and a laterally compressed body and tail. Body scales mostly small and homogeneous, with widely spaced transverse rows of larger scales.

**SUBSPECIES:** The Eastern Water Dragon (*Intellagama lesueurii lesueurii*) is yellowish brown to greyish brown with up to seven broad dark dorsal bands, a prominent broad dark stripe from eye to side of neck, and a reddish flush on chest and abdomen, most intense on mature males.

The Gippsland Water Dragon (*Intellagama lesueurii howittii*) is olive green to bluish green with dark dorsal markings reduced to absent and no dark bar behind eye. Ventral surfaces are olive green and throats of males are blackish with streaks and blotches of yellow, orange and occasionally blue.

**KEY CHARACTERS:** Differs from all other dragons in having a laterally compressed tail.

**DISTRIBUTION AND ECOLOGY:** Margins of water courses such as creeks and rivers, drainage channels, lakes and ornamental ponds in eastern Australia. The Gippsland Water Dragon extends from eastern Vic, north to the Shoalhaven River area, NSW, where it is replaced northwards to about Cooktown, Qld by the Eastern Water Dragon. Introduced populations of both subspecies occur along parts of the Yarra River near Melbourne. Terrestrial and arboreal, basking and foraging on river banks, on rocks and in riverside vegetation. They are also semi-aquatic, readily taking to water if disturbed. These excellent swimmers, propel themselves forward by undulating their laterally compressed bodies and tails with limbs held to their sides. They may remain submerged for many minutes.

**BIOLOGY:** These are familiar lizards in Canberra, Sydney, Brisbane and many other urban areas, where they become habituated in parks and gardens. There, fat lazy lizards bask beside ornamental ponds and accept titbits from passers-by. However, away from human habitation, Water Dragons are shy, difficult to approach and quick to retreat.

Genetic and morphological studies in Brisbane have identified several inner city populations narrowly separated by unsuitable urban landscapes, which have begun to diverge. They exhibit genetic diversity and differences in size and limb length. The lizards demonstrate that evolution, under certain circumstances, can proceed at a much more rapid pace than generally acknowledged (Littleford-Colquhoun *et al*, 2017).

this page
*Intellagama lesueurii howittii.*
Croajingalong National Park. Vic
photo D. Paul

opposite page
right:
*Intellagama lesueurii howittii.*
Croajingolong National Park. Vic
photo D. Paul

left:
*Intellagama lesueurii* habitat.
St Ives, NSW
photo S. K. Wilson

Clutches usually comprise 6–12, but up to 17 eggs, laid in burrows high on creek banks. Males are significantly larger than females, with higher crests and brighter colours. They generally occupy a stretch of river bank with several adult females. Juveniles tend to stay well away from adults, possibly to avoid cannibalism. Male combat has been recorded, with both individuals lying chin to chin before grasping each other's jaws. They are broadly omnivorous, taking arthropods, the occasional small vertebrate, flowers, fruits and soft foliage. Captive individuals have been recorded to live for more than 20 and up to 26 years (Wilson, 2012).

*Intellagama lesueurii lesueurii.*
Brisbane River, Qld
photo S. K. Wilson

*Intellagama lesueurii lesueurii.*
Mt Cootha, Qld
photo S. K. Wilson

*Lophognathus horneri.*
Dampier Peninsula, WA
photo S. K. Wilson

# LOPHOGNATHUS

## TA-TA LIZARDS

*Lophognathus* Gray, 1842

**DESCRIPTION:** Two moderately large species (SVL 86 mm). Robust with long heads, powerful, moderately long limbs and long tail. Head wide with extensive covering of spinose scales. Dorsal scales weakly heterogeneous with enlarged vertebral and dorsolateral rows including enlarged nuchal crest of flattened spinose scales and erectable vertebral crest. Keels of dorsal scales aligned parallel to midline. Typically grey or brown to black with prominent pale dorsolateral stripes, usually straight-edged but occasionally deeply notched along inner edges to form elongate blotches, and usually narrowly discontinuous with wide pale stripe along upper and lower jaws. Well-defined pale stripe between eye and ear. Males most sharply patterned in black and white. Femoral pores 2–7; preanal pores 3–6.

**KEY CHARACTERS:** Differs from *Tropicagama* in having heterogeneous dorsal scales with keels parallel to midline (versus homogeneous scales with keels converging towards midline), more than 2 preanal pores and by having the pale dorsolateral stripe usually disjunct from stripe along jaw (versus broadly continuous). Differs from *Amphibolurus* in having dorsal scales less strongly heterogeneous.

**DISTRIBUTION AND ECOLOGY:** Semi-arboreal, occurring in dry tropical woodlands across northern and north-western Australia. Often more common along wooded margins of water courses. Common garden and parkland lizards in some northern towns.

**BIOLOGY:** They perch on vegetation, tree branches and rocks. The common name Ta-ta Lizard, applied commonly across much of the range, is derived from the frequent hand-waving (circumduction) behaviour.

**COMMENTS:** Genetic work has confirmed that *Lophognathus* is unrelated to *Tropicagama*. (Melville *et al*, 2011).

# LOPHOGNATHUS

## GILBERT'S DRAGON, TA-TA LIZARD

*Lophognathus gilberti* Gray 1842

**DESCRIPTION:** SVL 86 mm. Robust with long head, powerful, moderately long limbs and long tail. Dorsal scales heterogenous with vertebral and dorsolateral rows of weak to prominent enlarged, strongly keeled scales. Shades of grey to brown with prominent pale dorsolateral stripes fading at hips, a broad white stripe on the upper and lower lips, extending along full extent of jaw, and a white, grey or yellow stripe from behind the eye to the top of the ear, bordered above and below by a row of dark scales. Femoral pores 2–7; preanal pores 3–6.

**KEY CHARACTERS:** Similar to *L. horneri*, with extensive distributional overlap. Differs in lacking a well-defined white spot on tympanum. Differs from *Amphibolurus centralis* in having a pale stripe between the eye and ear and far fewer spinose dorsal scales. Differs from *Tropicagama temporalis* by having a pale stripe between the eye and ear, heterogeneous dorsolateral scales and more than 2 preanal pores. Differs from *Gowidon longirostris* by having shorter limbs and tail, lacking 1-3 white spots on a black background behind the ear and having fewer than10 femoral pores. Juvenile *L. gilberti* may be mistaken for some *Diporiphora* species but differ in having a broad white lip stripe extending the length of the jaw.

**DISTRIBUTION AND ECOLOGY:** Woodlands and tropical savannahs, particularly along waterways, of far northern Australia from coastal Kimberley, extending north of Kununurra to north of Katherine, NT, and through Arnhem Land to western Qld.

**BIOLOGY:** Territorial lizards, where males actively defend areas and can be observed head bobbing, hand waving, tail flicking and raising erectable nuchal crests. When active they are most commonly found in full shade, close to vegetation, on a raised perch such as a branch, rock or log (Thompson & Thompson, 2001). They commonly use bipedal movement, are known to be capable swimmers and have been observed diving to avoid capture. The hand waving (circumduction) is sometimes performed when no other lizards are in sight, such as just after a short bout of activity. It has led to the popular name of 'Ta-ta Lizards'. Females are reproductively active in the build-up and the wet season, probably laying two clutches of eggs a season. Clutches of 4–8 eggs have been recorded (James & Shine, 1988). Some of this biological data may be derived from the recently described *L. horneri* Melville *et al*, 2018

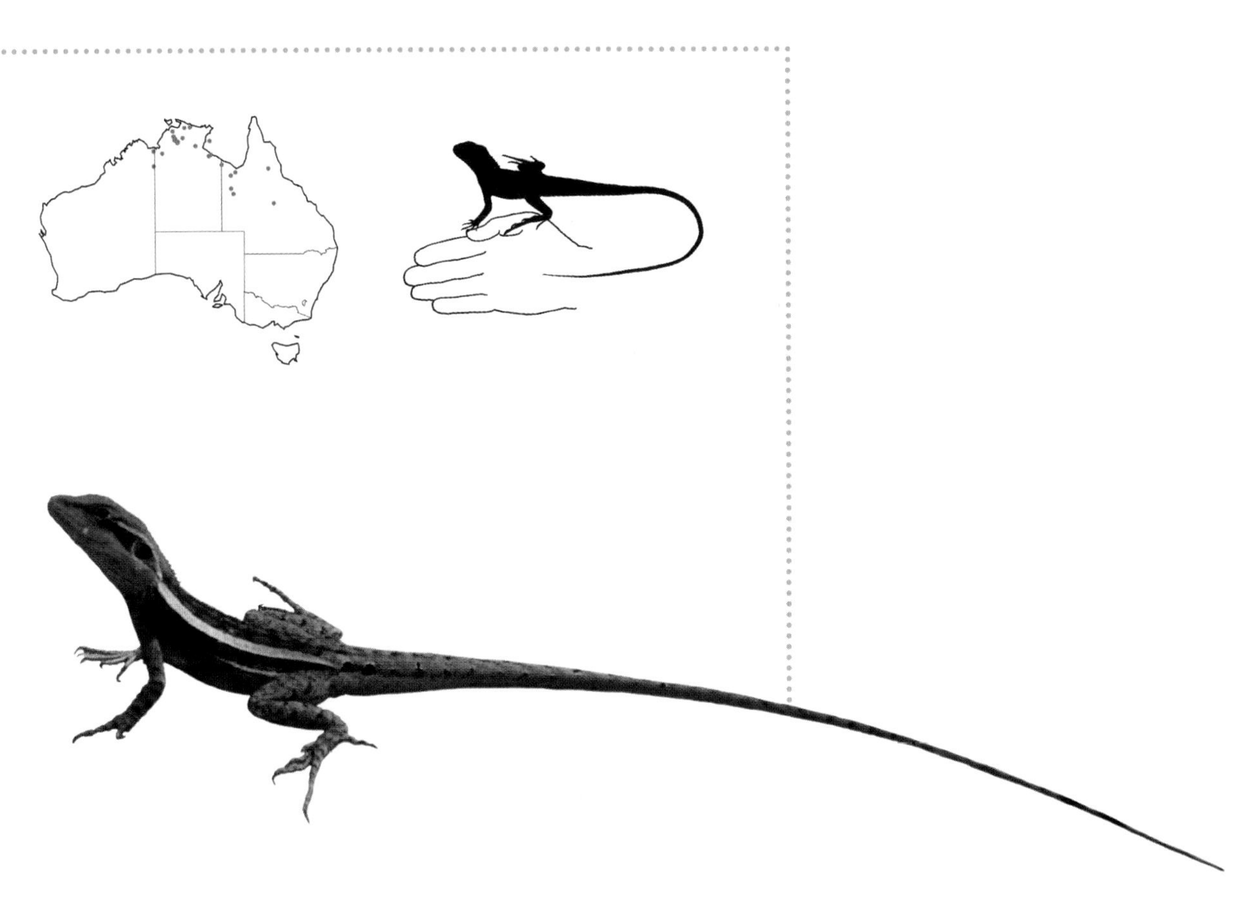

this page
*Lophognathus gilberti.*
Katherine, NT
photo R, Glor

opposite page
right:
*Lophognathus gilberti.*
Katherine, NT
photo R, Glor

left:
*Lophognathus gilberti*
habitat. Katherine, NT.
photo R, Glor

## HORNER'S DRAGON

*Lophognathus horneri* Melville, Ritchie, Chapple, Glor & Schulte, 2018

**DESCRIPTION:** SVL 86 mm. Robust with long head, powerful, moderately long limbs and long tail. Dorsal scales heterogeneous with weak to prominent vertebral and dorsolateral rows of enlarged, strongly keeled scales. Shades of brown, grey to black with prominent pale dorsolateral stripes, sometimes notched on their inner edges, fading at hips. Broad white stripe on the upper and lower lips extends along the full extent of the jaw, and a white, grey or yellow pale stripe extends from behind the eye to the top of the ear, bordered above and below by a row of dark scales. Tympanum has a distinct white spot, wholly surrounded or bordered by black. Breeding males are black, with sharply contrasting white dorsolateral and jaw stripes. Femoral pores 2–8; preanal pores 3–6.

**KEY CHARACTERS:** Similar to *L. gilberti*, with extensive distributional overlap. Differs in having a well-defined white spot on tympanum. Differs from *Amphibolurus centralis* in having less spinose dorsal scales, a well-defined white spot on tympanum and a pale stripe between the eye and ear. Differs from *Tropicagama temporalis* by having a pale stripe between the eye and ear, heterogeneous dorsolateral scales and more than 2 preanal pores. Differs from *Gowidon longirostris* by having shorter limbs and tail, lacking 1–3 white spots on a black background behind the ear and having fewer than10 femoral pores. Juvenile *L. horneri* may be mistaken for *Diporiphora* species but differ in having a broad white lip stripe extending the length of the jaw.

**DISTRIBUTION AND ECOLOGY:** Arid and semi-arid eucalypt woodlands and tropical savannahs of northern Australia. Has been recorded as far south as the Davenport Ranges, NT, and in WA it extends into the Kimberley and down to the northern Pilbara coast, southwest to Coral Bay and offshore islands.

**BIOLOGY:** Similar to *L. gilberti* in behaviour, including display rituals involving head bobbing, arm waving and raising an erectable nuchal crest.

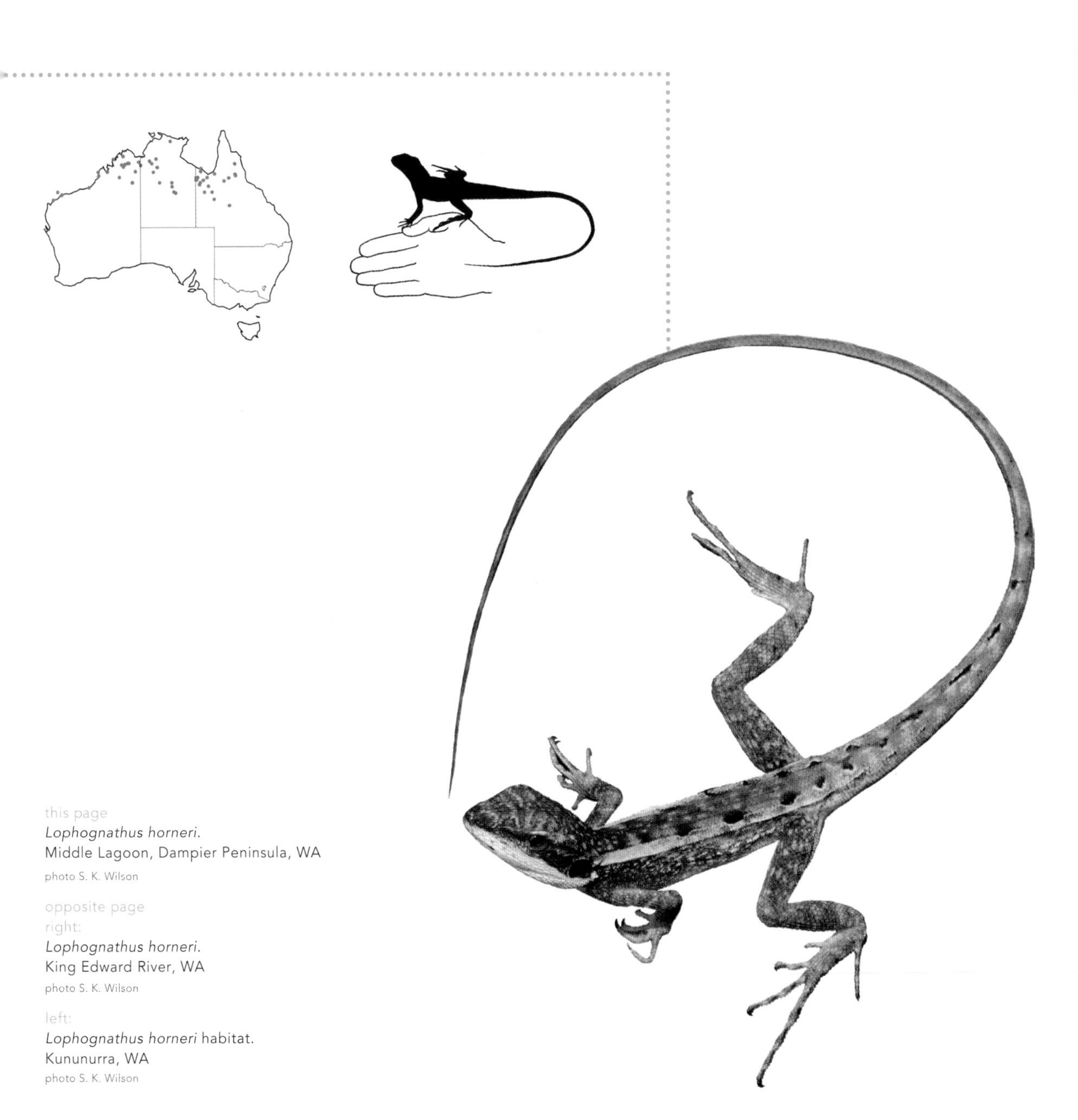

this page
*Lophognathus horneri.*
Middle Lagoon, Dampier Peninsula, WA
photo S. K. Wilson

opposite page
right:
*Lophognathus horneri.*
King Edward River, WA
photo S. K. Wilson

left:
*Lophognathus horneri* habitat.
Kununurra, WA
photo S. K. Wilson

*Lophognathus horneri*.
Mt Isa, Qld.
photo S. K. Wilson

opposite page:
*Lophognathus horneri*. Mature male.
Sandfire, WA.
photo M Bruton

*Lophosaurus boydii.*
*Mossman Gorge, Qld.*
photo S. K. Wilson

# LOPHOSAURUS

## RAINFOREST DRAGONS

*Genus Lophosaurus* Fitzinger, 1843

**DESCRIPTION:** Two Australian species (another in New Guinea) with very angular brows, strongly laterally compressed bodies, moderately long tails, very long slender limbs and large nuchal and vertebral crests of spines. Body scales are heterogeneous, with small keeled body scales and scattered larger keeled to spinose scales. Femoral and preanal pores are absent.

**KEY CHARACTERS:** Distinguished from all other dragon genera in having strongly laterally compressed body with tail that is round in cross-section, an extendable dewlap and conspicuous nuchal and vertebral crests comprising a raised ridge with large compressed spines.

**DISTRIBUTION AND ECOLOGY:** These dragons inhabit rainforests and adjacent wet sclerophyll forests in widely separated areas. The Boyd's Forest Dragon (*Lophosaurus boydii*) of Queensland's Wet Tropics is separated by a gap of at least one thousand kilometres from the Southern Angle-headed Dragon (*L. spinipes*), which occurs in the subtropical forests of southern Queensland and northern New South Wales.

**BIOLOGY:** These arboreal dragons are slow moving and cryptic, resting motionless on trunks and sapling stems. Their angular, serrated outline probably helps conceal them against the variegated forest backdrop. And rather than betraying themselves by dashing for cover if approached, they tend to slide discreetly from view around the other side of the trunk.

They are primarily arthropod feeders, which probably consume small vertebrates opportunistically. From their elevated vantage points they can locate moving prey on the forest floor and return to their perching sites after feeding. They take little or no vegetation, and any herbage consumed is probably taken incidentally with prey-capture.

Rainforest dragons are unusual among Australian dragons in the way they manage their temperatures. Other dragons actively bask in the sun and maintain optimal temperatures by either moving to or away from the heat source, adopting postures that enhance or reduce the amount of heat that reaches their bodies, and even selecting colours that absorb or reflect heat. In contrast rainforest dragons operate a temperature management system known as 'thermo-conforming'. Rather than seeking the shafts of sunlight that penetrate their forest habitat they simply allow their body temperatures to rise and fall with the ambient air temperatures.

## BOYD'S FOREST DRAGON

*Lophosaurus boydii* Macleay, 1884

**DESCRIPTION:** SVL 150 mm. The high nuchal and vertebral crests are discontinuous above the forelimbs. There are enlarged, very prominent plate-like scales on the cheeks, and the deep gular pouch has a row of very large tooth-spines along its leading edge. Rich purplish brown to greenish brown with a large black flush on the shoulder, bisected by a broad cream to white bar.

**KEY CHARACTERS:** The high, discontinuous crest and spiny gular pouch distinguish this species from all other Australian dragons.

**DISTRIBUTION AND ECOLOGY:** Rainforests and adjacent wet sclerophyll forests of the Wet Tropics in north Qld, occupying lowland and upland forests from the Paluma Range north to Shipton's Flat near Cooktown.

**BIOLOGY:** Up to six eggs are laid, typically in clearings such as a tree-fall or road verge. However lowland populations often lay in closed forest, presumably because of the higher ambient temperatures. See genus for general behavioural notes.

**NOTES:** Boyd's Forest Dragons are common across a wide area, much of it remote and protected under World Heritage legislation.

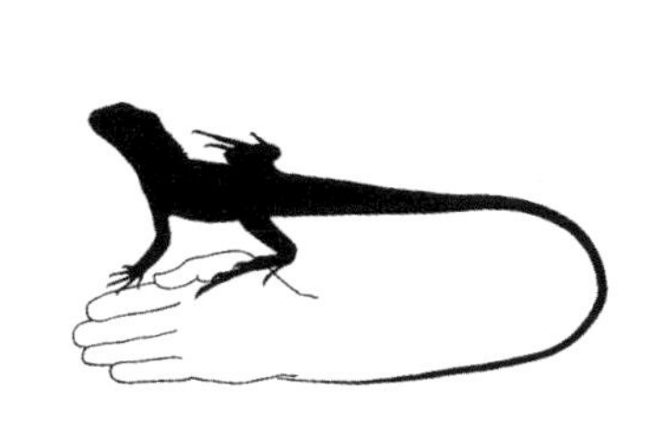

this page
*Lophosaurus boydii.*
Mossman Gorge, Qld
photo J. Melville

opposite page
right:
*Lophosaurus boydii.*
Mossman Gorge, Qld
photo S. K. Wilson

centre:
*Lophosaurus boydii.*
Mossman Gorge, Qld
photo S. K. Wilson

left:
*Lophosaurus boydii* habitat.
Cape Tribulation, Qld
photo S. K. Wilson

## SOUTHERN ANGLE-HEADED DRAGON

*Lophosaurus spinipes* Dumeril and Dumeril, 1851

**DESCRIPTION:** SVL 110 mm. The high nuchal and vertebral crests are continuous, with no disruption above the forelimbs. Unlike the Boyd's Forest Dragon, it lacks enlarged, very prominent plate-like scales on the cheeks, and the gular pouch has no spines along its leading edge. The limbs are covered with enlarged scattered spiny scales. Brown to greenish brown, sometimes with irregular blotches or variegations.

**KEY CHARACTERS:** Differs from the Water Dragon (*Intellagama lesueurii*) in having the tail round in cross-section (versus laterally compressed).

**DISTRIBUTION AND ECOLOGY:** Subtropical rainforests and adjacent wet sclerophyll forests of south-eastern Qld and north-eastern NSW, from the Gympie area south to about Gosford, NSW.

**BIOLOGY:** Eggs are typically laid in clearings such as tree-falls, walking tracks and road verges. Laying has been recorded between November and January. See genus for general behavioural notes.

**NOTES:** Southern Angle-headed Dragons are common across a wide area, much of it remote and protected in national parks.

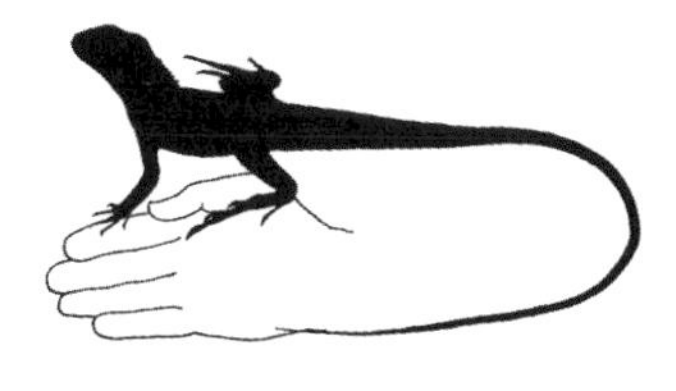

this page
*Lophosaurus spinipes.*
Mt Glorious, Qld
photo S. K. Wilson

opposite page
right:
*Lophosaurus spinipes.*
Mt Glorious, Qld
photo S. K. Wilson

left:
*Lophosaurus spinipes* habitat.
Lamington National Park, Qld
photo S. K. Wilson

*Moloch horridus*.
Hay River, NT
photo S. K. Wilson

# MOLOCH

## THORNY DEVIL

*Moloch* Gray, 1841

*Moloch horridus* Gray, 1841

## THORNY DEVIL

*Moloch* Gray,1841; *Moloch horridus* Gray, 1841

**DESCRIPTION:** Sole member of genus. SVL 110 mm. Extremely distinctive, with a dumpy body, short limbs and tail covered with an armoury of large thorny spines. There is a horn-like pair of spines above each eye and a similar pair on a bulbous hump on nape. Tail is thick, blunt and shorter than SVL and digits are short and thick. Colour pattern is a striking, sharply contrasting combination of black, white, yellow and ochre blotches and narrow white vertebral and dorsolateral stripes. These are extremely effective at breaking the animal's outline against dappled shade.

**KEY CHARACTERS:** The Thorny Devil cannot be confused with any other lizard.

**DISTRIBUTION AND ECOLOGY:** Sandy flats and dunes of central and western deserts, from the west coast to far western Qld. It is usually associated with spinifex on red sand, but on the west coast and south-western interior it occupies semi-arid complex heathlands. This slow-moving lizard has a characteristic jerky gait like a mechanical toy with the tail up-curved, and if disturbed it may freeze in mid-stride with a leg poised in mid-air. It is normally encountered on roads, but can be located by following the distinctive tracks on soft sand.

**BIOLOGY:** Thorny Devils are most active in early spring, moving up to 200–300 m daily, with males moving much further than females (Pianka *et al*, 1998). Egg-laying has been recorded between September and November in the Great Victoria Desert. Excavation of a nest hole took three days and clutches of 6–7 eggs were reported (Pianka *et al*, 1996). Considerable time is also spent disguising the nest hole, with stems, leaves and other vegetation raked carefully over the area (A. Sundholm, pers. com.).

The diet is highly specialised, comprising small black ants, primarily of the genus *Iridomyrmex*. Thorny Devils position themselves above an ant trail, feeding with repeated dips of the head as the ants pass under their nose. They may consume 750 or so ants per meal (Withers & Dickman, 1995). Each lizard visits its own regular defecation site where large numbers of scats containing embedded ant remains lie scattered.

Between the large spines are fine, granular scales, each with an overlapping shelf around its edge enclosing microscopic channels. If a Thorny Devil contacts water, the effect is like blotting paper, drawing water into a complex capillary system between the scales. It reaches corners of the mouth, apparently allowing the lizard to drink. This unusual strategy

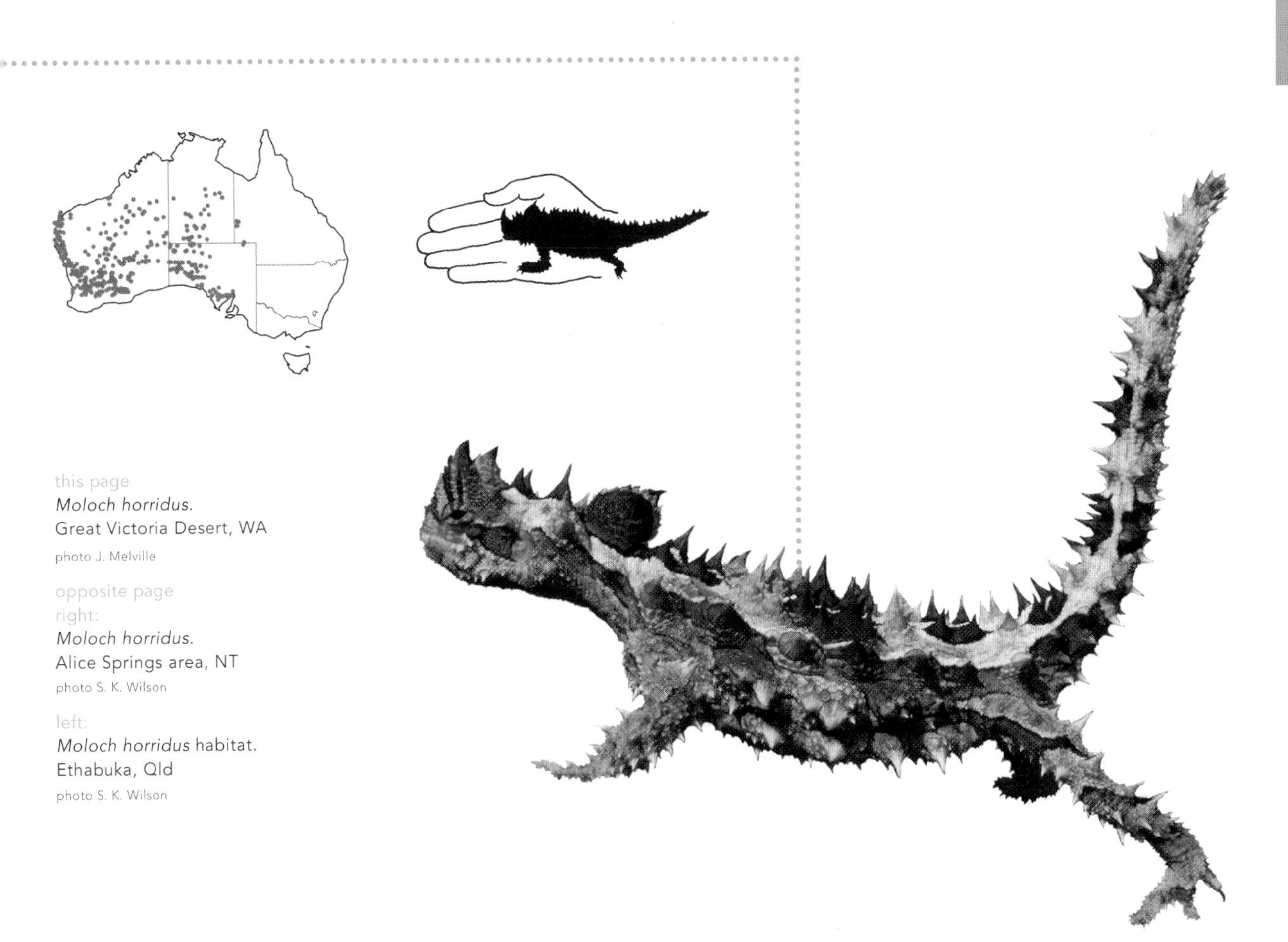

this page
*Moloch horridus.*
Great Victoria Desert, WA
photo J. Melville

opposite page
right:
*Moloch horridus.*
Alice Springs area, NT
photo S. K. Wilson

left:
*Moloch horridus* habitat.
Ethabuka, Qld
photo S. K. Wilson

is mirrored by the un-related, strikingly similar Texas Horned Lizard (*Phrynosoma cornutum*: Family Phrynosomatidae) from North America, as a method for the capture, transport and drinking of water from sporadic rainfall (Sherbrooke *et al*, 2007).

Debate continues about the hump on the neck, and it is often regarded as a decoy. If harassed a lizard may lower the head to present the hump, like a 'false head'. Yet there is little evidence of lizards with damaged humps, so discussion continues.

**NOTES:** Genetic and chromosomal work provides strong evidence that this species along with the Chameleon Dragon (*Chelosania brunnea*) form two separate, ancient lineages in the arid zone and are independent from the large, more recent radiation of dragons (Schulte *et al*, 2003; Hutchinson & Hutchinson, 2011).

*Moloch horridus.*
Goldfields Region, WA
photo R. Glor

*Pogona vitticeps*.
Longreach area, Qld
photo S. K. Wilson

# POGONA

## BEARDED DRAGONS

*Pogona* Storr, 1982

**DESCRIPTION:** Six described species of moderate to very large dragons (up to SVL 250 mm) with weakly to strongly dorsally depressed bodies and relatively short limbs and tails. Body scales heterogeneous; small scales mixed with large spines on back, rows of long prominent spines along flanks, a row across rear of head, and spines on the corner of jaw. These sometimes extend across the throat, and align with an erectable pouch supported by hyoid bones to form a 'beard'. Femoral and preanal pores are present and widely spaced.

**KEY CHARACTERS:** Distinguished from all other dragon genera in having a row of spines across the back of the head, and further from the Frilled Lizard (*Chlamydosaurus kingii*) in usually having a spiny erectable 'beard' under the throat (versus a thin erectable scaly frill that almost completely encircles the head.

**DISTRIBUTION AND ECOLOGY:** Bearded dragons are widespread across most of continental Australia. Habitats vary according to species. Most prefer woodlands and dry forests, but the Downs and Nullarbor Bearded Dragons (*P. henrylawsoni* and *P. nullarbor*) occupy very open grasslands or shrublands. Bearded dragons select elevated perching sites such as logs, stumps, and rocks. In some areas they are conspicuous along roadsides as they perch on fence posts. Sadly they are often also obvious as roadkill.

**BIOLOGY:** Bearded dragons are large members of the family agamidae, and in keeping with a general trend among most lizard families, there is a dietary shift from insectivory to omnivory and herbivory with increasing size. Young bearded dragons, with a body size approximating that of the adults of many other dragon species, feed largely on invertebrates. However, adults of the largest species consume a high percentage of vegetable matter in the form of flowers, fruits and soft foliage.

The erectable pouch or 'beard' is presented with a gaping pink or yellow mouth when an animal perceives a threat and during territorial disputes. The impressive display is often combined with dorsally flattening the body and turning the broadest aspect to the intruder. The combined effect is to appear large and fierce. The beard and defensive postures are best developed and employed in the largest species, the Common and Central Bearded Dragons (*Pogona barbata* and *P. vitticeps*). They are often called 'frillies', because of a superficial resemblance to the true Frill-necked Lizard (*Chlamydosaurus kingii*).

# POGONA

## EASTERN BEARDED DRAGON

*Pogona barbata* Cuvier, 1829

**DESCRIPTION:** SVL 266 mm. Head long and moderately narrow when viewed from above. Spines across rear of head arranged in a backward-curving arc, and spines across throat very well developed. Several rows of very long slender spines along flanks. Bands of enlarged scales around base of tail, separated by several bands of smaller scales. 'Beard' large and squarish when erected, and mouth-lining bright yellow. Shades of grey to yellowish brown with two rows of roughly circular paler blotches on body. Capable of colour change, adopting dark colours when basking and paler colours when optimum temperatures have been achieved. Mature males develop dark throats, particularly during threat displays towards potential danger or rivals. Femoral and preanal pores total 10–30.

**KEY CHARACTERS:** Differs from Central Bearded Dragon (*P. vitticeps*) by head moderately narrow (versus very broad and triangular), spines across base of head curving backwards (versus a straight transverse row), presence of bands of enlarged scales around base of tail (versus absent), squarish (versus round) beard, and yellow (versus usually pink) mouth-lining.

**DISTRIBUTION AND ECOLOGY:** Woodlands, wooded farmlands and some urban regions of eastern Australia, extending west to the Eyre Peninsula, SA. and north to the Mareeba area, Qld. Urban populations are common in parklands, golf courses, cemeteries and gardens in cities such as Canberra, Sydney and Brisbane. Juveniles are more terrestrial than adults, spending much of their time at or near ground level, while adults are more arboreal (Wotherspoon & Burgin, 2011).

**BIOLOGY:** Average summer field body temperatures of approx. 35°C have been recorded in Qld (Schauble & Grigg, 1998). Juveniles are mainly insectivorous, taking large active prey suggesting they are active hunters. Adults are more omnivorous. They take smaller arthropod prey and tend to be sit-and-wait predators. They also eat a variety of plant material such as flowers, soft foliage and fruits. However males, particularly the largest ones, lean most towards herbivory (Wotherspoon & Burgin, 2016). This suggests that females continue to require more protein than males throughout maturity. They need to produce eggs while males defend territories using mainly bluff rather than active aggression. Disputes are largely ritualised posturing but they may degenerate into combat, grasping each other's jaws and sometimes inflicting damage to the snout, teeth and even breaking jaws. A study of northern populations found that males are reproductively active year-round, with only a brief period of regression in late summer. Females store sperm in small storage pockets along the reproductive tract during the breeding season (spring-summer) and produce 2–3 large clutches

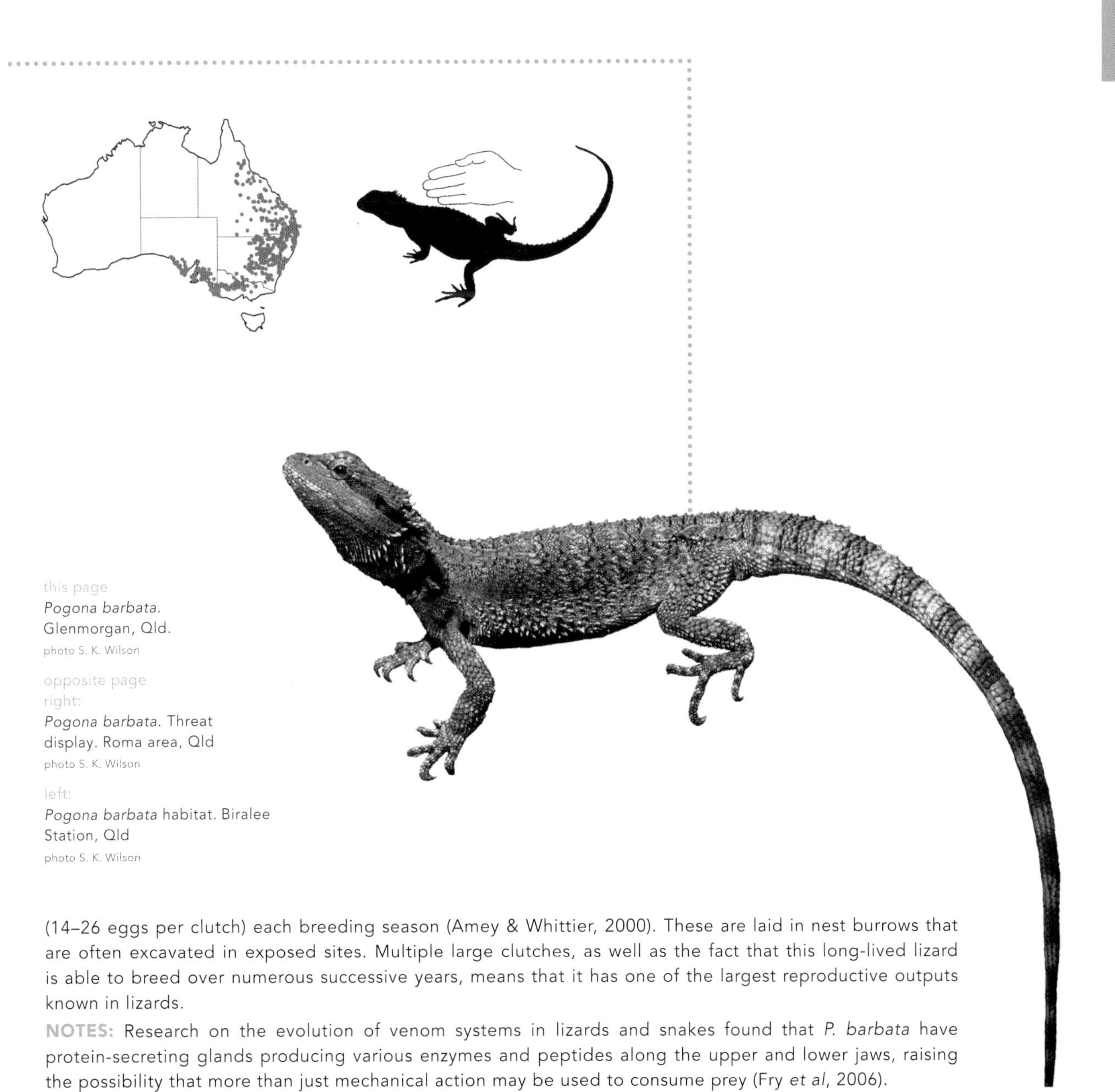

this page
*Pogona barbata*.
Glenmorgan, Qld.
photo S. K. Wilson

opposite page
right:
*Pogona barbata*. Threat display. Roma area, Qld
photo S. K. Wilson

left:
*Pogona barbata* habitat. Biralee Station, Qld
photo S. K. Wilson

(14–26 eggs per clutch) each breeding season (Amey & Whittier, 2000). These are laid in nest burrows that are often excavated in exposed sites. Multiple large clutches, as well as the fact that this long-lived lizard is able to breed over numerous successive years, means that it has one of the largest reproductive outputs known in lizards.

**NOTES:** Research on the evolution of venom systems in lizards and snakes found that *P. barbata* have protein-secreting glands producing various enzymes and peptides along the upper and lower jaws, raising the possibility that more than just mechanical action may be used to consume prey (Fry *et al*, 2006).

## POGONA

### DOWNS BEARDED DRAGON

*Pogona henrylawsoni* Wells and Wellington, 1985

**DESCRIPTION:** SVL 148 mm. Head bulbous, with short limbs and tail. Spines across rear of head short and conical, arranged in a straight line to very weakly backward-curving arc, spines across throat very poorly developed to absent, a few enlarged spines at rear corner of jaw, a line from behind eye to above ear, a cluster on shoulders and 1–4 rows of acute spines along flanks. 'Beard' very small and rounded when erected, and mouth-lining pink. Grey to yellowish brown with two rows of roughly circular paler blotches on body.

**KEY CHARACTERS:** Differs from other bearded dragon in having a greatly reduced beard, little or no indication of spines across throat, bulbous head and very short appendages.

**DISTRIBUTION AND ECOLOGY:** Mitchell grass plains on cracking clay in interior of Qld. Unlike other bearded dragons it inhabits an essentially treeless environment, utilising shrubs and grasses, earth clods and low occasional stones as perching sites and soil cracks as shelter sites.

**BIOLOGY:** Poorly known but assumed to be similar to other bearded dragons. Up to 20 eggs recorded in a clutch.

**NOTES:** One of the most frequently kept Australian lizards, in both Australia and overseas herpetoculture.

opposite page left:
*Pogona henrylawsoni* habitat.
Winton district, Qld
photo S. K. Wilson

opposite page right:
*Pogona henrylawsoni.*
Julia Creek, Qld
photo S. K. Wilson

below:
*Pogona henrylawsoni.*
Richmond area, Qld
photo S. K. Wilson

*Pogona henrylawsoni*. Longreach area, Qld
photo S. K. Wilson

top:
*Pogona minor mitchelli.*
Hillside Station, WA.
photo J. de Jong

bottom:
*Pogona minor minima.*
East Wallabi Island, WA
photo R. Browne-Cooper

## NULLARBOR BEARDED DRAGON

*Pogona nullarbor* Badham, 1976

**DESCRIPTION:** SVL 140 mm Short snout. Robust, strongly depressed body. Spines across rear of head arranged in backward curving arc and spines small but continuous across throat forming a moderately weak 'beard'. From 3–7 rows of long spines along flanks. Interior of mouth yellow. Grey to orange-brown with 6–7 narrow irregular pale bands on back and neck, and sometimes a paired series of pale blotches between nape and hips.

**KEY CHARACTERS:** Differs from the Western Bearded Dragon (*P. m. minor*) in having a fully developed 'beard' with spines extending across the throat (versus poorly developed with spines confined to the rear edges of jaw), and several rows of spines along flanks (versus one row).

**DISTRIBUTION AND ECOLOGY:** Restricted to shrublands of the Nullarbor Plain in south-eastern WA and south-western SA, occupying an essentially treeless habitat of chenopod shrubs and areas of limestone rocks. It shares with the Downs Bearded Dragon, *Pogona henrylawsoni*, a preference for areas without trees. All other bearded dragon species occur largely in woodlands or open forests.

**BIOLOGY:** Clutches of 14 eggs are recorded (Smith & Schwaner, 1981). Otherwise poorly known but similar to other bearded dragons. Selects shrubs and rocks as elevated basking sites.

opposite page left:
*Pogona nullarbor* habitat.
Border Village area, SA
photo A. Samuel

opposite page right:
*Pogona nullarbor*.
Naretha area, WA
photo S. Ford

below:
*Pogona nullarbor*.
Cocklebiddy, WA
photo B. Schembri

## CENTRAL BEARDED DRAGON

*Pogona vitticeps* Ahl, 1926

**DESCRIPTION:** SVL 250 mm. Head broad and triangular when viewed from above. Spines across rear of head arranged in a straight transverse line, and spines across throat very well developed. Several rows of very long slender spines along flanks. Scales around base of tail of roughly uniform size. 'Beard' large and rounded when erected with spines extending fully across throat, and mouth-lining usually pink, occasionally yellow. Shades of grey to yellowish brown or brick red with two rows of roughly circular paler blotches on body. 'Beard' of adult male may be black, particularly during displays to rivals.

**KEY CHARACTERS:** Differs from Eastern Bearded Dragon (*P. barbata*) by head very broad and triangular (versus moderately narrow), spines across base of head in a straight transverse row (versus curving backwards), round (versus squarish) 'beard', spines along flanks arranged regularly (versus very irregularly) and lacking enlarged bands of scales on base of tail (versus bands of large scales separated by several bands of smaller scales). Differs from the Western Bearded Dragon (*P. m. minor*) in having a broad (versus narrow) head, a fully developed 'beard' with spines extending across the throat (versus poorly developed with spines confined to the rear edges of jaw), and several rows of spines along flanks (versus one row).

**DISTRIBUTION AND ECOLOGY:** Woodlands and shrublands of the eastern interior, inland to central Australia. May be locally abundant, including in disturbed pastoral areas. A common sight in the outback perching on fence posts, stumps and roadside debris.

**BIOLOGY:** Ecologically very similar to the Eastern Bearded Dragon (*P. barbata*), probably sharing the same sexual and age related dietary differences. Average active body temperature of approx. 33°C has been recorded (Melville & Schulte, 2001). Up to 35 eggs are recorded in a clutch (Greer, 1989). It has genetically determined sex, but can shift to temperature determined sex under extremely high incubation temperatures. This shift has been documented in wild populations in Qld and northern NSW (Holleley *et al*, 2015).

**NOTES:** The genome of *P. vitticeps* has been sequenced (Georges *et al*, 2015) and this species has been the focus of extensive research on the evolution of sex chromosomes. One of the most frequently kept Australian lizards, in both Australia and overseas herpetoculture.

opposite page left:
*Pogona vitticeps* habitat.
Ethabuka, Qld
photo S. K. Wilson

opposite page right:
*Pogona vitticeps.*
Wagon Flat Bore, Vic
photo S. K. Wilson

below:
*Pogona vitticeps.*
Ballera area, Qld
photo S. K. Wilson

top:
*Pogona vitticeps.*
Alice Springs, NT
photo J. Melville

bottom:
*Pogona vitticeps.*
Bokhara Ruins, NT
photo S. K. Wilson

opposite page:
*Pogona vitticeps.*
Alice Springs, NT
photo J. Melville

*Rankinia diemensis.*
Anglesea, Vic
photo S. K. Wilson

# RANKINIA

## MOUNTAIN DRAGON

*Rankinia Wells and Wellington,*
*1984 Rankinia diemensis Gray, 1841*

# RANKINIA

## MOUNTAIN DRAGON

*Rankinia* Wells and Wellington, 1984; *Rankinia diemensis* Gray, 1841

**DESCRIPTION:** Sole member of genus. SVL 82 mm. Small and robust with relatively short limbs and tail. Dorsal scalation heterogeneous, comprising small scales, with much larger spinose scales scattered over back and flanks, and arranged in paravertebral, dorsolateral and lateral rows. Another prominent row of spinose scales extends along each side of the tail-base. Tympanum is exposed. Grey to reddish brown vertebral stripe extends from nape to hips, edged by much broader pale dorsal stripe with straight outer edges, and deeply serrated inner edges highlighted in black, creating a prominent zigzagging pattern.

**KEY CHARACTERS:** Differs from *Tympanocryptis* in having an exposed tympanum (versus covered by scales). Differs from the Jacky Lizard (*Amphibolurus muricatus*) in having a pink mouth lining (versus yellow) and further from both in having lateral rows of enlarged spines on tail-base.

**DISTRIBUTION AND ECOLOGY:** Heath and open forest, often on sandy soils, in temperate areas of south-eastern Australia. Distribution is patchy. Disjunct populations include Tas. (where it is the only agamid and the world's most southerly) and Flinders Island in Bass Strait; Grampians National Park in western Vic.; Anglesea on the south coast; and along the Great Dividing Range from eastern Vic to near Tamworth, NSW. It is wholly terrestrial, foraging in open spaces between vegetation and seldom perching on elevated sites. It is an alert lizard, but not particularly swift relative to other dragons.

**BIOLOGY:** One of the few Australian agamid lizards where females are larger on average than males. This difference in adult body size has been attributed to a reproductive advantage where larger females are able to have larger clutches (Stuart-Smith *et al*, 2008). On mainland Australia, egg-laying has been recorded in November and December, with clutches of 2–7 eggs deposited in burrows excavated in open areas, hatching in March or April (Kent, 1987). In Tas., *R. diemensis* has a winter torpor that lasts seven months, with males emerging from this torpor in September, before females. In these southern latitudes, females breed from October to January, producing up to two clutches of 2–11 eggs (Stuart-Smith *et al*, 2005). Females are able to store sperm over the breeding season and a second clutch of eggs can be laid 5 weeks after the first.

opposite page left:
*Rankinia diemensis* habitat.
Anglesea, Vic
photo J. Melville

opposite page right:
*Rankinia diemensis.*
Anglesea, Vic
photo S. K. Wilson

below:
*Rankinia diemensis.*
ACT
photo J. de Jong

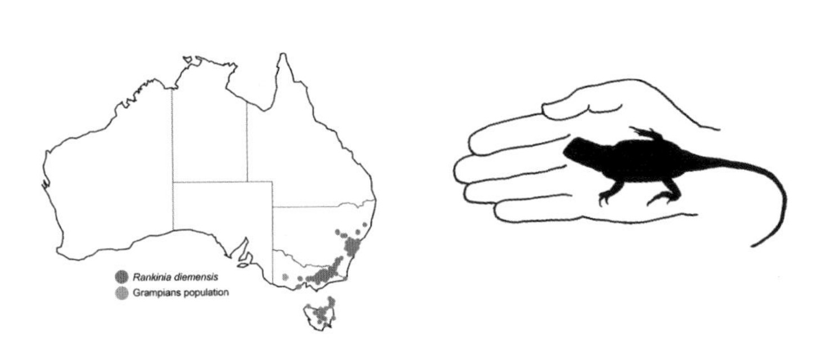

**NOTES:** Genetic evidence indicates the Grampians population is more closely related to those from Tas. and Bass Strait than to the geographically closer population from Anglesea. And based on genetic divergence and isolation, the Grampians and Anglesea populations are sufficiently distinct to warrant definition as Evolutionary Significant Units. Each is at risk and worthy of specific conservation efforts (Ng *et al*, 2014). The Grampians population (highlighted in blue on map) is on the Advisory List of Threatened Vertebrate Fauna in Victoria as Critically Endangered.

*Tropicagama temporalis.*
Saibai Island, Qld
photo S. K. Wilson

# TROPICAGAMA

## SWAMPLANDS LASHTAIL; NORTHERN WATER DRAGON

*Tropicagama* Melville, Ritchie, Chapple, Glor & Schulte, 2018 *Tropicagama temporalis* Günther, 1867

## TROPICAGAMA

### SWAMPLANDS LASHTAIL; NORTHERN WATER DRAGON

*Tropicagama* Melville, Ritchie, Chapple, Glor & Schulte, 2018;
*Tropicagama temporalis* Günther, 1867

**DESCRIPTION:** Sole member of genus. SVL 120 mm. Moderately slender with very long-limbs, long slender tail, and prominent erectable nuchal crest. Head relatively narrow. Dorsal scales uniform, with keels converging posteriorly toward midline. Brown to black with prominent pale dorsolateral stripes broadly continuous with wide pale stripe along upper and lower jaws. No well-defined pale stripe between eye and ear. Males very sharply and simply patterned; head, chin, throat and neck uniformly dark grey or black, and pale stripes strongly contrasting white. Females and juveniles often less uniform in colour with brown and black patterning on top of the head and flecks of grey, brown or black on ventral surfaces. One or more broad dark lateral bands commonly present across back and shoulders. Femoral pores 1-6; preanal pores 1–3.

**KEY CHARACTERS**: Superficially similar to and has extensive distributional overlap with Gilbert's Dragon (*Lophognathus. gilberti*) but differs by having 3 or less preanal pores (versus more than 3), uniform dorsal scales, having the prominent pale dorsolateral stripe broadly continuous with stripe along jaw (versus disjunct) and no well-defined pale stripe between eye and ear (versus present).

**DISTRIBUTION AND ECOLOGY:** Semi-arboreal, occurring in dry tropical woodland habitats, particularly associated with coastal pandanus and paperbark watercourses. Far northern Australian coastal regions in NT, western Cape York, Qld, and possibly northern coastal WA. Also occurs on Torres Strait islands, Indonesian islands close to Australian waters, and southern New Guinea. A common inhabitant in Darwin's suburbs and urban bushland.

**BIOLOGY:** Often seen perched on vegetation, tree branches and rocks, hand-waving and head bobbing when disturbed. Although it has been observed to be active throughout the year in the suburbs of Darwin, it probably has period of inactivity during the late dry season elsewhere.

**NOTES:** Genetic work has confirmed that *T. temporalis* is unrelated to *Gowidon*, *Amphibolurus* and *Lophognathus* species (Melville *et al*, 2011).

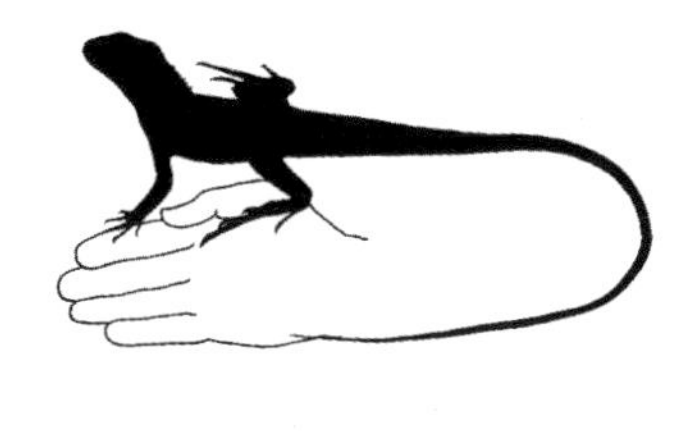

this page
*Tropicagama temporalis.*
Casuarina, Darwin, NT
photo R. Coupland

opposite page
right:
*Tropicagama temporalis.*
Casuarina, Darwin, NT
photo R. Coupland

left:
*Tropicagama temporalis*
habitat. Saibai Island, Qld
photo S. K. Wilson

*Tympanocryptis macra.*
Top Springs, NT
photo R. Glor

# TYMPANOCRYPTIS

## EARLESS DRAGONS

*Tympanocryptis* Peters, 1863

**DESCRIPTION:** Seventeen described species of small dragons (up to SVL 74 mm) with compact bodies and relatively short limbs and tails. Body scales heterogeneous; small scales mixed with enlarged spines over back. Tympanum completely covered by scales. Femoral and preanal pores are present or absent, but never more than 2 of each.

**KEY CHARACTERS:** Distinguished from nearly all other dragon genera by a combination of small, compact build, completely hidden tympanum and heterogeneous dorsal scalation featuring enlarged spinose scales. Differs from *Ctenophorus butleri* and *C. parviceps* in lacking an enlarged row of spinose scales on each side of the tail-base.

**DISTRIBUTION AND ECOLOGY:** Earless dragons occur in dry, open, largely treeless habitats ranging from the harshest, most featureless stony deserts to eastern temperate and tropical grasslands. All species are terrestrial though some frequently perch on low objects such as rocks, small timber, shrubs and tussocks.

**BIOLOGY:** Desert species, are extremely tolerant to high temperatures, thanks largely to selective posturing. They are often the last reptiles remaining exposed as the temperature climbs. They can be seen perching on road-side stones with bodies held high off the substrate, sometimes even erect on the hind limbs, angled into the sun to minimise direct exposure. White undersides help reflect heat radiating from below. A juvenile Smooth-snouted Earless Dragon (*T. intima*) has been observed basking on a dark metal chain in full sun, when the shade temperature measured 43 degrees.

Some western arid adapted earless dragons (*T. cephalus* and related species) have evolved remarkable pebble mimicry. Round heads and bodies, reddish colours with little or no pattern and short legs tucked close to the sides combine to render the lizards virtually invisible as they nestle among small gibber stones. In contrast, eastern grassland species employ a different, yet effective camouflage technique. Bold contrasting patterns of pale dorsal and vertebral stripes overlying dark transverse bars effectively break up the lizards' shapes and outlines against a variegated background of tussock grass foliage.

Earless dragons are probably exclusively arthropod feeders, with ants comprising a significant portion of the diet.

Most species are secure across large tracts of land. However some of the eastern grassland inhabitants are among Australia's most threatened reptiles due to severe loss and fragmentation of their habitat. The Victorian Grassland Earless Dragon (*Tympanocryptis pinguicolla*) was once common west of Melbourne but is believed to have been extirpated from Victoria since the 1960s. Three other grassland earless dragons have highly restricted distributions in ACT (*T. lineata*) and NSW (*T. mccartneyi* and *T. osbornei*). And the Condamine Earless Dragon (*T. condaminensis*) is restricted to narrow road verges and the cotton, sorghum and maize crops of Queensland's Darling Downs. Its future rests largely on a continuation of current 'minimum till' agricultural practices.

## CENTRALIAN EARLESS DRAGON

*Tympanocryptis centralis* Sternfeld, 1924

**DESCRIPTION:** SVL 61 mm. Narrow neck, and head scales slightly heterogeneous and weakly keeled. Dorsal surface strongly heterogeneous; small smooth and weakly keeled scales, with scattered enlarged, strongly keeled scales. Ventral and gular scales weakly keeled. Reddish brown to light brown with pale stripes; a continuous vertebral, two narrower broken dorsolaterals and a lateral. Vertebral and dorsolateral lines merge at back legs and continue onto the tail as wide white stripe with dark brown blotches at regular intervals. Dorsal series of 4–5 dark blotches extend between dorsolateral and vertebral stripes, alternating with 4–5 pale bands that often disrupt dorsolateral stripes. Limbs and tail strongly patterned with alternating pale and dark banding, and white stripe on anterior edge of lower back leg. Solid pale band between eyes, weak in some individuals. Females tend to have less distinct patterning than males. Femoral pores 0; preanal pores 2.

**KEY CHARACTERS:** Not known to overlap with other species but *T. intima* and *T. tetraporaphora*, occur nearby. Differs from *T. tetraporaphora* in lacking femoral pores (versus one on each side), and from *T. intima* in having white dorsal and lateral stripes (versus absent).

**DISTRIBUTION AND ECOLOGY:** Occurs on central ranges of southern NT, north to western Barkly Tableland and south to the SA border. Records include Tennant Creek, West Macdonnell Ranges, Kings Canyon NP and Kata Tjuta. Occurs on arid stony flats or loams with scattered Triodia. Average active body temperatures of 35°C have been recorded (Melville & Schulte, 2001).

**BIOLOGY:** See genus *Tympanocryptis*.

this page:
*Tympanocryptis centralis*. Tennant Creek area, NT
photo S. K. Wilson

opposite page
right:
*Tympanocryptis centralis*. Kata Tjuta, NT
photo J. de Jong

left:
*Tympanocryptis centralis* habitat. Kata Tjuta, NT
photo S. Macdonald

*Tympanocryptis centralis.*
West MacDonnell Ranges, NT
photo J. Melville

*Tympanocryptis centralis.*
Kata Tjuta, NT
photo M. Hutchinson

# TYMPANOCRYPTIS

## COASTAL PEBBLE-MIMIC DRAGON

*Tympanocryptis cephalus* Günther, 1867

**DESCRIPTION:** SVL 64 mm. Rotund body, small bulbous head with angular snout, dorsoventrally compressed body, moderately gracile limbs and short tail. Snout convex when viewed from side, with nostril below canthus. Scales on snout have low keels. On dorsal surface of body, enlarged scales with raised spines are arranged in transverse rows of 5–7 scales. Weakly defined row of enlarged scales at anterior and dorsal edge of thigh, and scales on dorsal surface of thigh not aligned to form a ridge. Ventral scales have low keels. Brown to reddish brown with about 5 irregular dark blotches along midline and no indication of any longitudinal stripes. Femoral pores 0; preanal pores 2.

**KEY CHARACTERS:** Differs from *T. fortescuensis* and *T. diabolicus* in having longer transverse rows of enlarged dorsal scales (5–7 versus 2–5), enlarged scales along leading edge of thigh not forming a conspicuous ridge, and having dark dorsal blotches. Differs from *T. gigas* in having enlarged dorsal scales in transverse rows (versus scattered).

**DISTRIBUTION AND ECOLOGY:** Restricted to a coastal strip in Pilbara region, WA, in vicinity of Karratha and Roebourne, west to Mardie Pool, east to Balla Balla Creek, and inland about 40 kilometres. Presumed to occur on red rocky loams and clay soils with *Triodia* and Snakewood (*Acacia ziphophylla*).

**BIOLOGY:** One of a small group of accomplished pebble-mimic dragons, with the head and body closely resembling two pebbles. When crouched immobile among scattered stones, it takes a keen eye to see through the disguise.

**NOTES:** WA's pebble-mimic earless dragons were reviewed by Doughty *et al* (2015). In previous publications, the name *T. cephalus* was applied to a composite of species occurring in stony arid areas of WA. Following the review, *T. cephalus* is now restricted to a narrow area of about 5000 square kilometres.

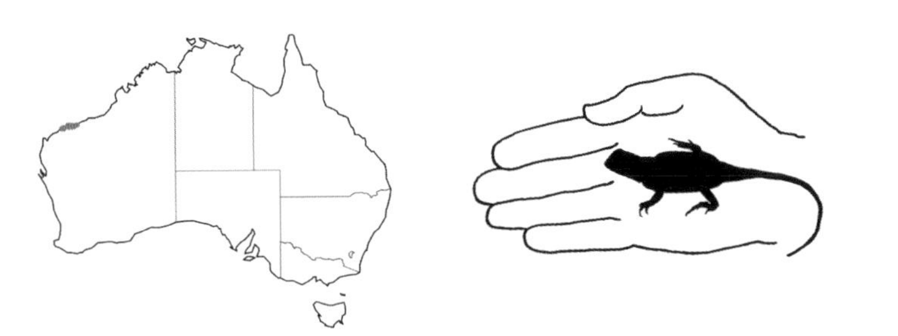

opposite page:
*Tympanocryptis cephalus* habitat.
Roebourne area, WA
photo G. Harold

below:
*Tympanocryptis cephalus*.
Karratha area, WA
photo G. Harold

## CONDAMINE EARLESS DRAGON

*Tympanocryptis condaminensis* Melville, Smith, Hobson, Hunjan & Shoo, 2014

**DESCRIPTION:** SVL 56 mm. Dorsal scales are heterogeneous with un-keeled to strongly keeled scales. More than 10 internasal scales. Well-developed dorsal, lateral and ventral body patterning and three well defined pale spots on dorsal surface of snout: one above each nostril and one at end of snout. Brown to grey with prominent pale vertebral, dorsolateral and usually continuous lateral stripes overlying a series of 4–5 broad dark bands that extend onto flanks. Distinct, irregular brown-black and white ventrolateral banding along flanks below lateral stripe, with pale colour predominating. Ventral patterning concentrated on throat and upper chest. Individuals with heavy ventral patterning usually have a narrow longitudinal pale median stripe along upper third of chest. A distinct narrow white stripe running along the posterior edge of the thigh, extending onto the base of the tail. Some individuals have red-pink colouration on throats and some have lemon yellow along flanks. Femoral pores 0; preanal pores 2.

**KEY CHARACTERS:** Distribution is not believed to overlap with any other *Tympanocrpytis* but is geographically close to Roma Earless Dragon (*T. wilsoni*). Differs in having a narrow white lateral stripe, lateral patterning of strongly contrasting irregular brown-black and white banding with more white than brown-black, and lacking femoral pores (versus one pore on each side).

**DISTRIBUTION AND ECOLOGY:** Restricted to the eastern Darling Downs in Qld, west to Dalby and east to the western outskirts of Toowoomba. Occurs on black cracking clays of the Condamine River floodplain in remnant native grasslands, croplands and roadside verges.

**BIOLOGY:** Strongly associated with black cracking clays, where it uses soil cracks for shelter. Land clearing and cropping practices have strongly impacted this species. With the exception of some narrow roadside verges of remnant native grasslands, the species is restricted to pastures of crops such as maize and sorghum. Individuals living in sorghum have been found to use litter as overnight refuges (Starr & Leung, 2006). Individuals with pink throats and yellow flanks, presumably breeding colours, have been recorded in October.

**NOTES:** Identified as species of high conservation priority, and listed as Endangered both in Qld and nationally. Listed as Endangered internationally on the IUCN Red List (Melville *et a.*, 2017a).

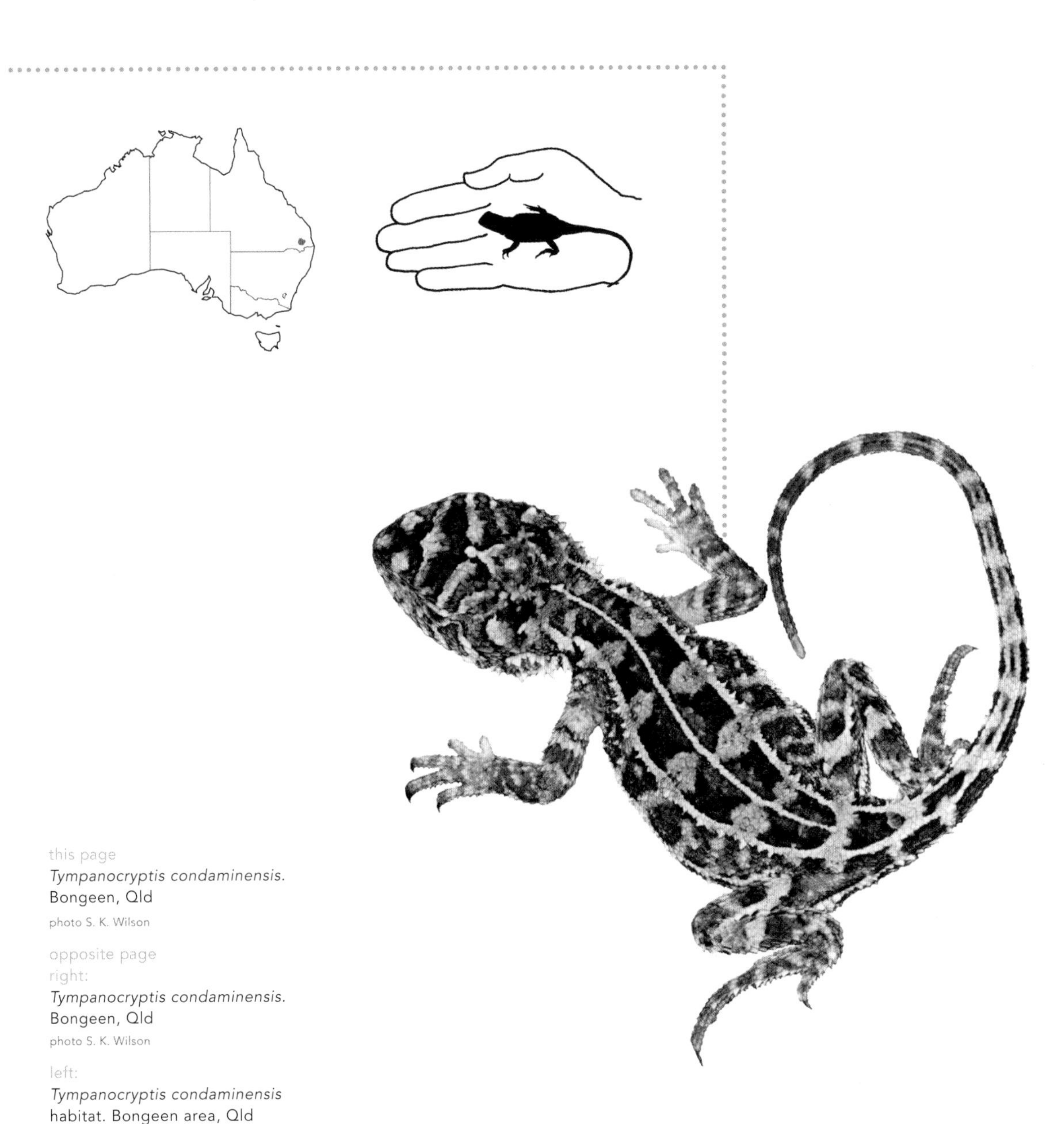

this page
*Tympanocryptis condaminensis.* Bongeen, Qld
photo S. K. Wilson

opposite page
right:
*Tympanocryptis condaminensis.* Bongeen, Qld
photo S. K. Wilson

left:
*Tympanocryptis condaminensis* habitat. Bongeen area, Qld
photo S. K. Wilson

## HAMERSLEY PEBBLE-MIMIC DRAGON

*Tympanocryptis diabolicus* Doughty, Kealley, Shoo & Melville, 2015

**DESCRIPTION:** SVL 60 mm. Very, rotund with small head, blunt snout, short neck and moderately short limbs. Snout straight or concave when viewed from side, with nostril below canthus, scales smooth to rugose with feeble keels and rostral scale 2–3 times wider than high. Dorsal scales include enlarged scales with raised spines, tending to be arranged in transverse rows of 2–5 scales. A row of enlarged scales along the anterior dorsal edge of thigh forms a conspicuous ridge, and keels on scales of upper arm and on dorsal surface of thigh are aligned. Ventral scales have low keels. Reddish-brown, often with a grey-black "wash" over head, body and limbs, three short, dark-edged white lines on nape and about four obscure dark blotches between nape and hips. No longitudinal stripes. Tail has about 8 pale bands with black anterior edges, merging on posterior half of tail to form a pale stripe. Femoral pores 0; preanal pores 2.

**KEY CHARACTERS:** Not known to overlap with any other *Tympanocrpytis* but occurs in close proximity to other Pilbara and Gascoyne species. Differs from *T. gigas* in having enlarged keeled scales in short transverse rows (versus scattered), from *T. cephalus* by having fewer enlarged scales in transverse rows (2–5 versus 5–7) and further from both by having enlarged scales along front of thigh forming a conspicuous ridge. Differs from *T. fortescuensis* by having scales on snout rugose with weak keels (versus with low keels), and colouration (reddish with midline dorsal blotches versus pale brown with little indication of blotches). Differs from *T. pseudosephos* by having low keels on ventral sales (versus smooth or slightly raised), and aligned keels on dorsal surfaces of upper arms and thighs (versus tending not to align).

**DISTRIBUTION AND ECOLOGY:** Hamersley Range area in the Pilbara region of WA, east to Mt Newman. Little known of ecology but recent surveys indicate preference for clayey substrates, with small rocks and pebbles strewn across the surface.

**BIOLOGY:** One of a small group of accomplished pebble-mimic dragons, with the head and body closely resembling two pebbles. When crouched immobile among scattered stones, it takes a keen eye to see through the disguise.

**NOTES:** WA's pebble-mimic earless dragons were reviewed by Doughty *et al* (2015). In previous publications, the name *T. cephalus* was applied to a composite of species occurring in stony arid areas of WA. In the review, *T diabolicus* was described as new and restricted to the Hamersley Range area. Genetic work reveals that *T. diabolicus* is closely related to *T. fortescuensis*.

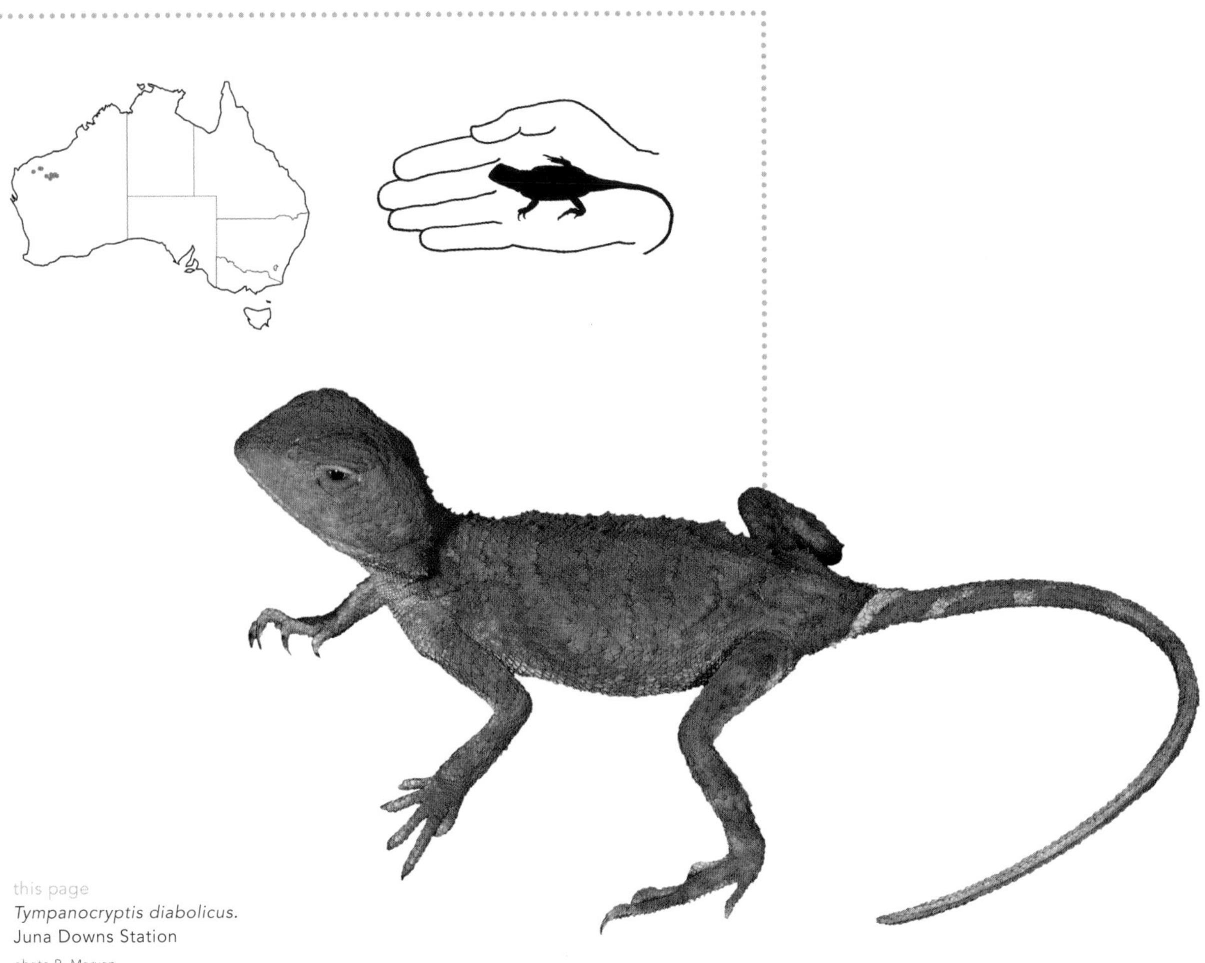

this page
*Tympanocryptis diabolicus.*
Juna Downs Station
photo B. Maryan

opposite page
right:
*Tympanocryptis diabolicus.*
Juna Downs Station, WA
photo R. Lloyd

left:
*Tympanocryptis diabolicus*
habitat. Hamersley Range, WA
photo B. Maryan

*Tympanocryptis diabolicus.*
Juna Downs Station
photo B. Maryan

## FORTESCUE PEBBLE-MIMIC DRAGON

*Tympanocryptis fortescuensis* Doughty, Kealley, Shoo & Melville, 2015

**DESCRIPTION:** SVL 65 mm. Rotund with small head, blunt snout, short neck and moderately-short limbs. Snout straight or convex when viewed from side, with nostril below canthus and slightly protruding mouth. Scales on snout smooth with low keels and rostral scale 2–3 times wider than high. Dorsal scales include enlarged scales with raised spines tending to be arranged in transverse rows of 2–5 scales. A row of enlarged scales along the anterior-dorsal edge of thigh forms a conspicuous ridge, and keels on scales of upper arm and dorsal surface of thigh are usually aligned. Ventral scales have low keels. Light reddish-brown, usually unpatterned but occasionally with subtle, barely discernible blotches. Nape dark brown with three short, dark-edged white lines. Limbs weakly banded, and tail has about 10 pale bands with black anterior edges, usually merging on posterior half to form a pale stripe. Femoral pores 0; preanal pores 2.

**KEY CHARACTERS:** Not known to overlap with any other *Tympanocrpytis* but mostly likely to be confused with nearby *T. cephalus* and *T. diabolicus*. Differs from *T. cephalus* in having enlarged dorsal scales in short transverse rows in rows of 2–5 scales (versus 5–7), conspicuous ridge of scales along anterior edge of thigh (versus weakly-defined row), light reddish-brown, usually uniform coloration or with barely discernible blotches (versus brown with dark blotches along midline) and scales on upper thigh usually aligned. Distinguished from *T. diabolicus* in having scales on snout with low keels (versus rugose with feeble keels) and light brown coloration that is usually uniform or with barely discernible blotches (versus rich reddish-brown, occasionally with blotches).

**DISTRIBUTION AND ECOLOGY:** Restricted to the Pilbara region, WA, around the Fortescue Marsh area and in cracking clay habitats in the Chichester Range, but may be more widely distributed north of the Fortescue River. Favours cracking (alluvial) clays with small scattered rocks and tussock grasses.

**BIOLOGY:** One of a small group of accomplished pebble-mimic dragons, with the head and body closely resembling two pebbles. When crouched immobile among scattered stones, it takes a keen eye to see through the disguise.

**NOTES:** WA's pebble-mimic earless dragons were reviewed by Doughty *et al* (2015). In previous publications, the name *T. cephalus* was applied to a composite of species occurring in stony arid areas of WA. In the review, *T. fortescuensis* was described as new and restricted to the Pilbara region. Genetic work reveals that *T. fortescuensis* is closely related to *T. diabolicus*.

opposite page:
*Tympanocryptis fortescuensis* habitat.
Pannawonnica area, WA
photo G. Harold

below:
*Tympanocryptis fortescuensis*.
Redmont area, WA
photo R. Lloyd

## GASCOYNE PEBBLE-MIMIC DRAGON

*Tympanocryptis gigas* Mitchell, 1948

**DESCRIPTION:** SVL 66 mm. Robust dragon with blunt snout, short neck and moderately gracile limbs. Snout convex when viewed from side, with nostril below canthus. Scales on snout are rugose with feeble keels and rostral scale twice as wide as high. Dorsal scales include scattered enlarged scales with raised spines, not arranged in transverse rows. Row of enlarged scales extend along anterior dorsal edge of thigh, scales on dorsal surface of thigh heterogeneous and not aligned and keels on scales on upper arm not aligned but those on lower arm aligned. Ventral scales smooth. Very pale brown to cream above with dark band and three pale stripes on nape and 1–2 large dark central blotches on body. Tail with alternating dark and pale bands. Femoral pores 0; preanal pores 2.

**KEY CHARACTERS:** Not known to overlap with any other *Tympanocryptis* but *T. diabolicus* and *T. pseudopsephos* occur nearby to the north in the Pilbara or east in the northern Goldfields region. Differs from *T. diabolicus* and *T. pseudopsephos* in having enlarged scales on dorsal surface scattered and not arranged in rows (versus enlarged scales in short transverse rows), and enlarged scales along front of thigh do not form a conspicuous ridge. Differs further from *T. pseudopsephos* by oblong (versus rotund) body shape and convex (versus concave) snout, and from *T. diabolicus* in having smooth ventral scales (versus with low keels).

**DISTRIBUTION AND ECOLOGY:** Arid stony areas of Gascoyne region, WA, including Yinnetharra and Williambury Stations. Habitat is poorly documented but at least one site includes a high proportion of pale quartz stones. The pale colouration of *T. gigas* may reflect a preference for such features.

**BIOLOGY:** One of a small group of accomplished pebble-mimic dragons, with the head and body closely resembling two pebbles. When crouched immobile among scattered stones, it takes a keen eye to see through the disguise.

**NOTES:** WA's pebble-mimic earless dragons were reviewed by Doughty *et al* (2015). In previous publications, the name *T. cephalus* was applied to a composite of species occurring in stony arid areas of WA. In the review, *T. gigas*, an old available name, was recognized as distinct, re-described and restricted to the Gascoyne area.

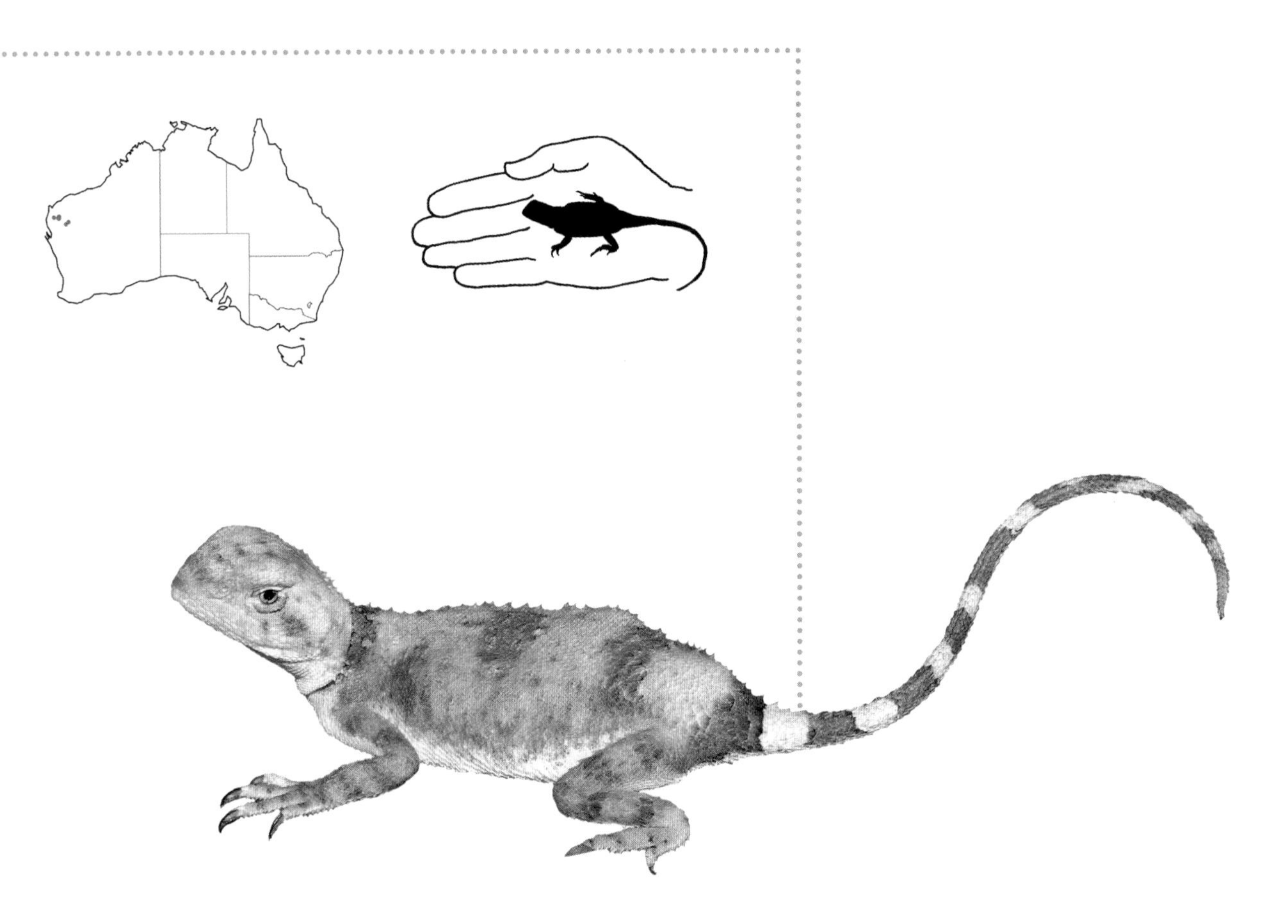

this page:
*Tympanocryptis gigas.*
Williambury Station, WA
photo B. Bush

opposite page
right:
*Tympanocryptis gigas.*
Williambury Station, WA
photo B. Bush

left:
*Tympanocryptis gigas* habitat.
Williambury Station, WA
photo B. Bush

## NULLARBOR EARLESS DRAGON

*Tympanocryptis houstoni* Storr, 1982

**DESCRIPTION:** SVL 62 mm. Indistinct neck as broad as back of head; moderately long limbs and tail. Head scales strongly keeled, scales along vertebral stripe weakly keeled and remaining dorsal scales strongly keeled. Dorsal scales include scattered slightly enlarged, keeled spinose scales. Ventral and gular scales smooth to weakly keeled. Reddish brown to grey-brown with broad pale grey vertebral stripe, and pale dorsolateral stripes 2–4 times narrower, overlaying 6 dark transverse bars. Usually a single narrow lateral stripe. Vertebral stripe becomes broken along tail, while dorsolateral stripes tend to continue down tail. Head strongly patterned with pale band between eyes and three pale spots on snout. Ventral surfaces white with scattered dark flecking or diffuse lines under head, and sometimes an orange flush in gular region. Femoral pores 0; preanal pores 2.

**KEY CHARACTERS:** At the eastern extent of its range, where *T. houstoni* nears the western extent of *T.* cf. *lineata*, it can be distinguished in having a pale vertebral stripe much broader than dorsolaterals.

**DISTRIBUTION AND ECOLOGY:** Nullabor Plain in WA and SA, occupying chenopod shrublands on clay soils and scattered stones.

**BIOLOGY:** Often seen perched on small stones beside the Eyre Highway.

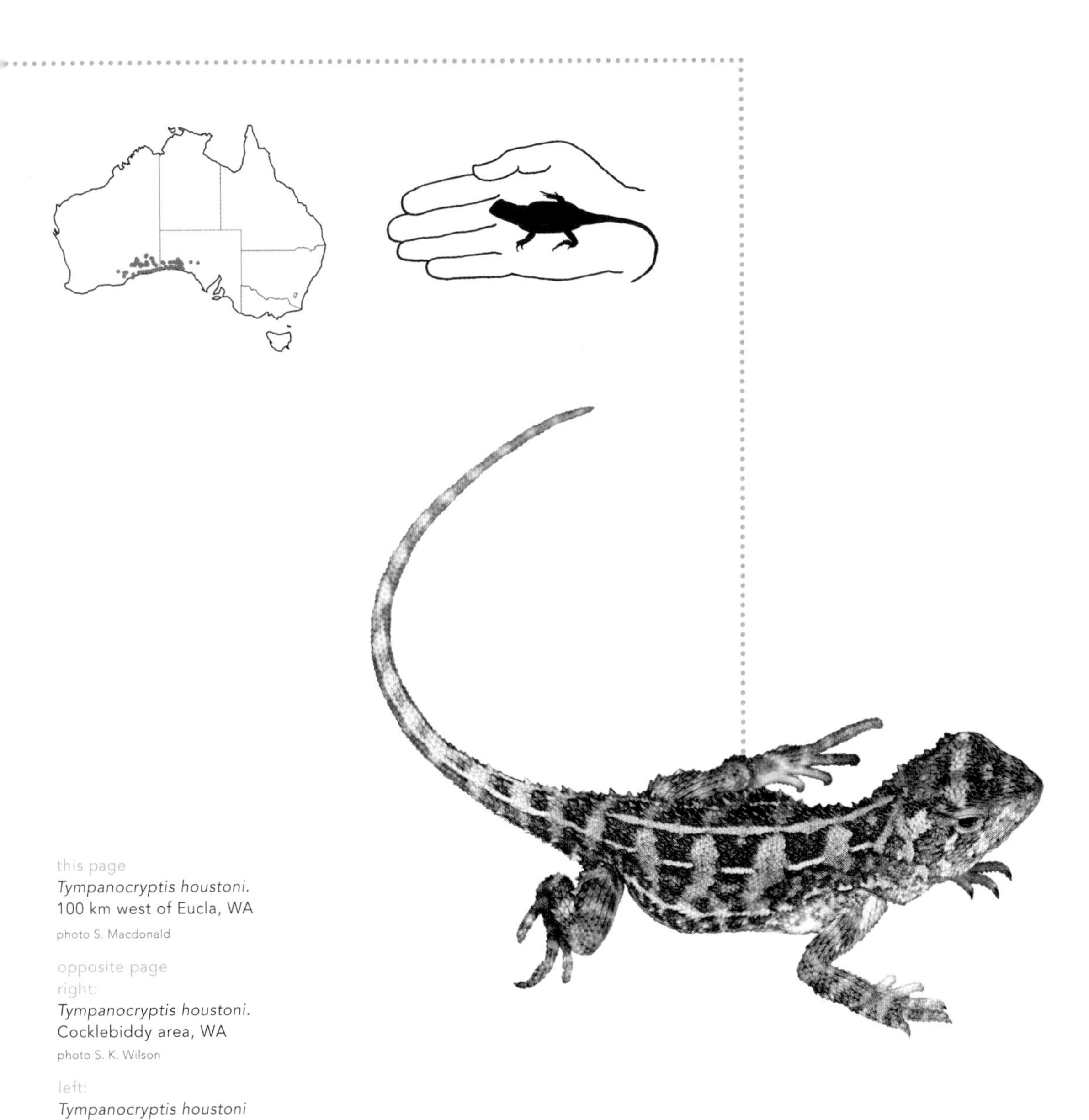

this page
*Tympanocryptis houstoni.*
100 km west of Eucla, WA
photo S. Macdonald

opposite page
right:
*Tympanocryptis houstoni.*
Cocklebiddy area, WA
photo S. K. Wilson

left:
*Tympanocryptis houstoni*
habitat. Nullarbor, WA
photo S. Zozaya

## SMOOTH-SNOUTED EARLESS DRAGON

*Tympanocryptis intima* Mitchell, 1948

**DESCRIPTION:** SVL 61 mm. Robust and dorsally flattened with rounded head and moderately short limbs. Dorsal surface covered with small un-keeled scales, with scattered enlarged tubercules with posteriorly projecting spines, tending to form irregular longitudinal rows. The enlarged tubercules are wider than long, with a raised edge along the posterior border and a single central spine protruding from the posterior edge. Limbs covered with weakly keeled scales, with interspersed enlarged tubercules. Ventral and gular scales smooth. Light grey to brown with about 4 dark brown blotches on either side of the vertebral line. Dorsal stripes usually absent, but sometimes an indication of broken narrow pale vertebral and dorsolateral stripes, often restricted to neck. Tail usually banded. Ventral surface white, occasionally flecked with brown in the gular and chest regions of males. Femoral pores 0; preanal pores 2.

**KEY CHARACTERS:** Often mistaken for species in the *T. cephalus group* of WA but these two lineages do not overlap in distribution. Differs from *T. tetraporaphora* and *T.* cf. *lineata* by usually lacking any longitudinal stripes on the dorsal surface. Differs further from *T. tetraporaphora* in having no femoral pores (versus one each side) and from *T.* cf. *lineata* in lacking a pale band between the eyes (versus present).

**DISTRIBUTION AND ECOLOGY:** Deserts of the Eyrean Basin in SA, NT, Qld and north-western NSW, extending north to the Mt Isa and Gulf region of Qld. Mainly inhabits gibber plains, with spinifex, tussock grasses or sparse shrubs, often on very featureless terrain.

**BIOLOGY:** Perches on stones and soil clods, often maintaining its stance in temperatures exceeding 40 degrees when all other reptile activity has ceased.

this page:
*Tympanocryptis intima.*
Birdsville area, Qld
photo S. K. Wilson

opposite page
right:
*Tympanocryptis intima.*
Eromanga area, Qld
photo S. K. Wilson

left:
*Tympanocryptis intima* habitat.
Bedourie - Birdsville road, Qld
photo S. K. Wilson

## LINED EARLESS DRAGON

*Tympanocryptis* c.f. *lineata* Peters, 1863

**DESCRIPTION:** SVL 58 mm. Indistinct neck as broad as back of head; moderately short limbs and tail. Head scales strongly keeled, and dorsal scales strongly keeled with scattered enlarged tubercules that are wider than long. Ventral and gular scales smooth to weakly keeled. Reddish brown to grey-brown with narrow pale vertebral stripe and broader, often broken, dorsolateral stripes overlaying 5 dark transverse dorsal bars. A pale lateral stripe, often ill-defined. Vertebral and dorsolateral stripes continue weakly onto the tail, fading into patterning. Head patterned with pale band between eyes and three pale spots on snout. Limbs and tail strongly patterned with alternating pale and dark banding. White ventral surfaces with dark brown or black flecking. Femoral pores 0; preanal pores 2.

**KEY CHARACTERS:** Possibly overlapping with *T. intima*, *T. tetraporaphora* and *T. houstoni* at edges of distribution. Differs *T. tetraporaphora* in having no femoral pores (versus one each side). Differs from *T. intima* in having prominent pale dorsal and lateral longitudinal stripes (versus little if any indication of stripes), having strongly keeled (versus smooth) head scales and indistinct neck as broad as back of head (versus narrow neck and distinct head). Differs from *T. houstoni* in having narrow vertebral stripe (versus vertebral stripe much broader than dorolaterals).

**DISTRIBUTION AND ECOLOGY:** South-central Australia, including central-eastern SA, north-western Vic and western NSW. Widespread in arid and semi-arid open habitats, including gibber plains, chenopod shrublands and mallee. In Victoria, mainly restricted to saltlake flats and surrounds.

**BIOLOGY:** Little is known. Active field body temperature of approximately 37°C have been recorded (Greer, 1989).

**NOTES:** Listed as threatened in Victoria as *T. lineata*. This species is currently under taxonomic revision, following a recent publication (Melville *et. al.*, 2019) that re-assigned the name *T. lineata* to the Canberra Grassland Earless Dragon (*T. lineata*).

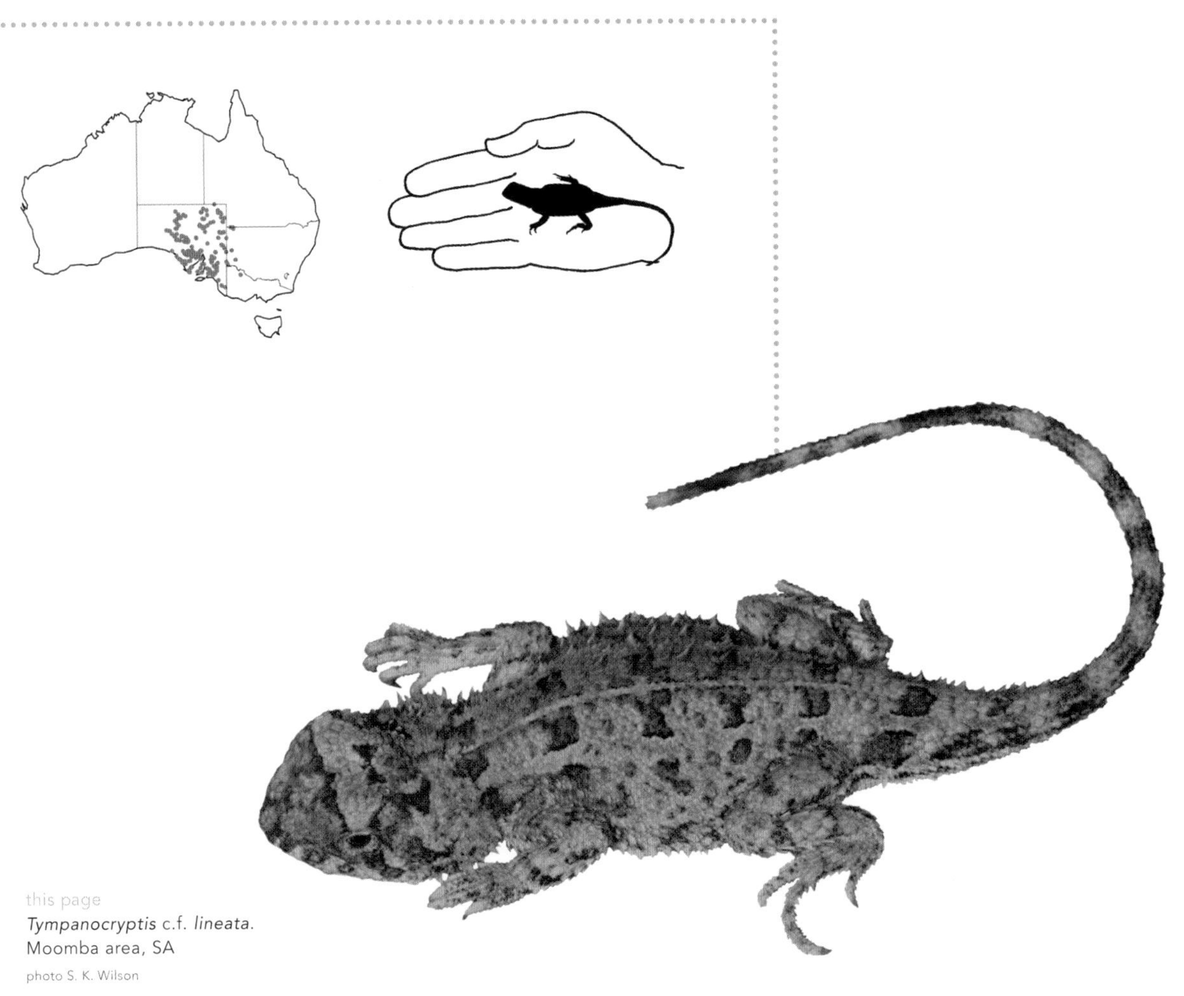

this page
*Tympanocryptis* c.f. *lineata*.
Moomba area, SA
photo S. K. Wilson

opposite page
right:
*Tympanocryptis* c.f. *lineata*.
Moomba area, SA
photo S. K. Wilson

left:
*Tympanocryptis* c.f. *lineata*
habitat. Innamincka area, SA
photo S. K. Wilson

## CANBERRA GRASSLAND EARLESS DRAGON

*Tympanocryptis lineata* Peters, 1863

**DESCRIPTION:** SVL 61 mm. Indistinct neck as broad as back of head; moderately short limbs and tail. Head scales strongly keeled, dorsal scales heterogenous and strongly keeled, with scattered enlarged tubercules with raised spines directed upwards and backwards. Ventral and gular scales smooth to weakly keeled. Reddish brown to grey-brown with prominent, pale vertebral and dorsolateral stripes, the dorsolaterals often slightly wider, overlaying about 6–7 dark transverse bands across back, and an ill-defined pale lateral stripe. Head patterned with indistinct pale band between eyes and three pale spots on snout. Both sexes exhibit an orange-pink flush on ventral surface. Some individuals, presumably breeding males, have the throat marked with a yellow flush and black marbling. Femoral pores absent; preanal pores 2.

**KEY CHARACTERS:** Not known to overlap with other species but differs from nearby *T. osbornei* in typically having fewer dark dorsal blotches on tail (7–11 versus 12-14), and from *T. mccartneyi* in having smooth (versus keeled) gular scales and no enlarged tubercles on thigh.

**DISTRIBUTION AND ECOLOGY:** Restricted to fragmented and diminishing temperate grasslands of ACT.

**BIOLOGY:** A secretive grassland specialist that is known to use spider burrows for shelter, with individuals having one or two home burrows. Studies have revealed small home ranges of up to about 5000 m2 (Stevens *et al*, 2010) and high site fidelity, which increases with the onset of winter, indicating the importance of burrows as over-winter refuge sites.

**NOTES:** Federally listed as Endangered and listed at state level as Endangered in ACT (as *T. pinguicolla* – a species now regarded as restricted to Vic). The name *T. lineata* has long been referred to a common and widespread species extending from north-western Vic and western NSW through much of SA. However the type specimen is from ACT so the name applies to this restricted species.

opposite page right:
*Tympanocryptis lineata.*
Canberra, ACT
photo S. K. Wilson

opposite page left:
*Tympanocryptis lineata* habitat.
Jerrabomberra Grassland Reserve, ACT
photo S. K. Wilson

below:
*Tympanocryptis lineata.*
Canberra, ACT
photo S. K. Wilson

## SAVANNAH EARLESS DRAGON

*Tympanocryptis macra* Storr, 1982

**DESCRIPTION:** SVL 52 mm. Tail and limbs relatively long for the genus. Dorsal patterning variable. Cream to brown with weak to strong pale vertebral and dorsolateral stripes of about equal width that extend down 2/3rds of the tail length, overlying 4-6 broken dark bands. Banding does not extend to ventral surfaces, but some individuals have dark pigmentation on throat and upper chest. Scales in armpits dark brown with scattered pale scales and flanks pale with scattered clusters of dark brown scales, creating a speckled appearance that can extend on to the belly in some individuals. Femoral pores 0; preanal pores 2.

**KEY CHARACTERS:** Distribution does not overlap any other *Tympanocrpytis*, although may potentially occupy the same regions as *T. uniformis* and *T. tetraporophora*. It differs from *T. tetraporophora*, in lacking femoral pores (versus 2 femoral pores). Differs from *T. uniformis* in having long limbs (versus short limbs) and having scattered enlarged tubercles on back (versus lacking enlarged tubercules).

**DISTRIBUTION AND ECOLOGY:** Tropical savannah grasslands and shrublands, mainly on cracking soils, on northern edges of arid zone from the Kimberley region in WA to mid-way across NT.

**BIOLOGY:** Behaviour poorly known but probably similar to *T. tetraporaphora*. Has been observed perching on litter, earth clods, or small rocks along road verges. Like *T. tetraporophora*, it has been seen to stand erect on hind limbs during hot weather. Known to use cracks in soil for shelter.

**COMMENTS:** *T. macra* was described as a subspecies of *T. lineata* but genetic work has provided strong evidence that this north-western earless dragon is unrelated to the earless dragons of southern Australia (Shoo *et al*, 2008). Thus, in this book we raise *T. macra* to full species status. Additionally, populations of *T. macra* in NT, particularly around the Wave Hill area, have been incorrectly identified as *T. uniformis* but based on both genetics and examination of the *T. uniformis* type specimen (held at the South Australian Museum), we can correctly attribute these NT populations to *T. macra*.

opposite page:
*Tympanocryptis macra* habitat.
Ellendale mine area, WA
photo G. Harold

below:
*Tympanocryptis macra.*
Ellendale mine area, WA
photo G. Harold

## BATHURST GRASSLAND EARLESS DRAGON

*Tympanocryptis mccartneyi*

Melville, Chaplin, Hutchinson, Sumner, Gruber, MacDonald and Sarre, 2019

**DESCRIPTION:** SVL 58 mm. Indistinct neck as broad as back of head; moderately short limbs and tail. Head scales strongly keeled, dorsal scales heterogeneous and strongly keeled, with scattered enlarged tubercules with raised spines directed upwards and backwards. Thigh scales heterogeneous with scattered enlarged tubercles. Ventral and gular scales keeled. Greyish-brown with prominent, pale vertebral, dorsolateral and lateral stripes overlying about 6 dark transverse bands across back. Head patterned with contrasting pale band between eyes and three pale spots on snout. Ventral surface with varying amount of dark mottling. Femoral pores absent; preanal pores 2.

**KEY CHARACTERS:** Not known to overlap with other species but differs from nearby *T. lineata* and *T. osbornei* in having scattered enlarged tubercles on thigh (versus absent) and keeled (versus smooth) gular scales.

**DISTRIBUTION AND ECOLOGY:** Known only from grasslands in the Bathurst area, NSW. Much of the known habitat is disturbed.

**BIOLOGY:** A secretive grassland specialist, known from only a few records. Presumed similar to other grassland earless dragons.

**COMMENTS:** This newly described species is probably of conservation concern as its habitat is small and fragmented, it is represented by only a few specimens, and related species of grassland earless dragons are known to have declined severely. There are no protected areas within its known distribution.

opposite page:
*Tympanocryptis mccartneyi* habitat.
Mt Panorama, Bathurst area, NSW
photo D Goldney

below:
*Tympanocryptis mccartneyi.*
Photographed c.1988.
Bathurst Plains, NSW
photo G. Waters

## MONARO GRASSLAND EARLESS DRAGON

*Tympanocryptis osbornei*

Melville, Chaplin, Hutchinson, Sumner, Gruber, MacDonald and Sarre, 2019

DESCRIPTION: SVL 58 mm. Indistinct neck as broad as back of head; moderately short limbs and tail. Head scales strongly keeled, dorsal scales heterogenous and strongly keeled, with scattered enlarged tubercules with raised spines directed upwards and backwards. Ventral and gular scales smooth. Reddish brown to grey-brown with prominent, pale vertebral, dorsolateral and lateral stripes. The vertebral is continuous and dorsolaterals are sometimes disrupted by about 6–7 dark transverse bands across back. Head patterned with indistinct pale band between eyes and three pale spots on snout. Both sexes exhibit an orange-pink flush on ventral surface. Some individuals have a bright yellow flush on throat. Femoral pores absent; preanal pores 2.

KEY CHARACTERS: Not known to overlap with other species but differs from nearby *T. lineata* in typically having more dark dorsal blotches on tail (12–14 versus 7–11), and from *T. mccartneyi* in having smooth (versus keeled) gular scales and no enlarged tubercles on thigh.

DISTRIBUTION AND ECOLOGY: Restricted to the Monaro Tablelands, NSW, in a small area north to Murrumbidgee River, south to Maclaughlin River, east to Monaro Hwy and west to Berridale. The area comprises fragmented native grasslands with scattered surface rocks.

BIOLOGY: A secretive grassland specialist that is known to use spider burrows and surface rocks for shelter.

COMMENTS: This newly described species is probably of conservation concern as its habitat is small and fragmented, studies demonstrate its numbers have declined, related species of grassland earless dragons are known to have declined severely and there are few protected areas within its known distribution.

opposite page left:
*Tympanocryptis osbornei* habitat.
Monaro, NSW
photo W. Osborne

opposite page right:
*Tympanocryptis osbornei*.
Monaro, NSW
photo S. K. Wilson

below:
*Tympanocryptis osbornei*. Cooma, NSW
photo S. Mahon

## FIVE-LINED EARLESS DRAGON

*Tympanocryptis pentalineata* Melville, Smith, Hobson, Hunjan & Shoo, 2014

DESCRIPTION: SVL 52 mm. Relatively slender with strongly keeled scales on top of head and body and weakly to strongly keeled scales on ventral surface. Dorsal scales include scattered enlarged, raised mucronate scales. Dorsal body pattern is distinctive. Brown-black with five longitudinal narrow grey or white stripes; narrow vertebral and dorsolaterals, and lateral stripes consisting of single rows of enlarged white scales bordered by smaller, darker scales. Three prominent broad, dark-brown to black bands present between the dorsolateral and lateral stripes. Throat and upper chest area sometimes faintly pigmented with black flecks. Femoral pores 2; preanal pores 2.

KEY CHARACTERS: Differs from *T. tetraporophora* and *T. intima* in having five longitudinal stripes, and further from *T. intima* by having two femoral pores (versus femoral pores absent).

DISTRIBUTION AND ECOLOGY: Only known from the one location, 50 km south-west of Normanton in the Gulf region of far northern Qld. Occurs on flat flood-plains, covered by grasses and low perennial shrubs.

BIOLOGY: Virtually nothing known about this recently described species but probably similar to *T. tetraporaphora*.

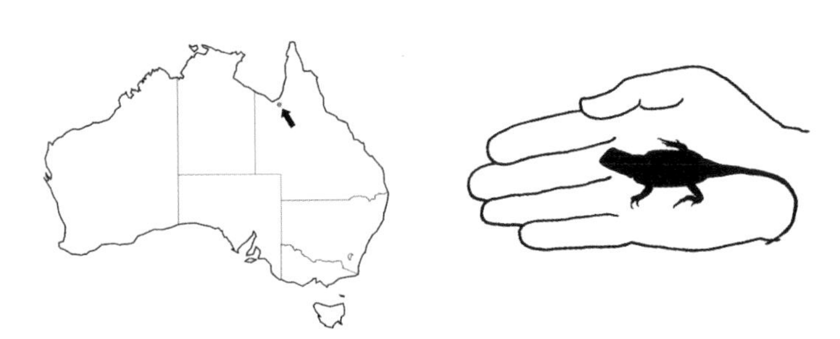

opposite page right:
*Tympanocryptis pentalineata.*
Normanton area, Qld
photo S. Macdonald

opposite page left:
*Tympanocryptis pentalineata* habitat.
Normanton area, Qld
photo S. Macdonald

below:
*Tympanocryptis pentalineata.*
Normanton area, Qld
photo S. Macdonald

## VICTORIAN GRASSLAND EARLESS DRAGON

*Tympanocryptis pinguicolla* Mitchell, 1948

**DESCRIPTION:** SVL 62 mm. Indistinct neck as broad as back of head with well-developed lateral skin-fold from jaw to gular fold; moderately short limbs and tail. Head scales strongly keeled, dorsal scales heterogeneous and strongly keeled, with numerous scattered, strongly enlarged spinous dorsal scales, at least twice the width of adjacent body scales, with the apex of spinose scales directed almost vertically. Reddish brown to grey-brown with prominent, narrow pale vertebral and dorsolateral stripes overlaying 5–7 dark transverse bands across back, and an ill-defined pale lateral stripe. An indistinct pale band between eyes. Some individuals, presumably breeding males, have the throat marked with a yellow flush and black marbling. Femoral pores absent; preanal pores 2.

**KEY CHARACTERS:** Does not overlap with other *Tympanocryptis* species.

**DISTRIBUTION AND ECOLOGY:** Restricted to grasslands on the basalt plains to the north and west of Melbourne. No specimens have been recorded since the 1960s, though there have been reported, unconfirmed sightings at Craigieburn, Merri Creek, Holden Flora Reserve and Little River west of Werribee between 1988 and 1990.

**BIOLOGY:** A secretive grassland specialist. An 1890 report by Lucas and Frost refers to the use of spider holes and loose basalt boulders as shelter sites. This concurs with the known habits of a close relative, the Canberra Grassland Earless Dragon (*T. lineata*) which has been well studied.

**COMMENTS:** Federally listed as Endangered. With no recorded specimens for about 50 years, despite extensive searches, and with all of its former grassland habitats highly modified for agriculture and urban development, it is feared that this may be the first mainland extinction of a native reptile. The last positive sightings were at the Little River Gorge, south-west of Melbourne.

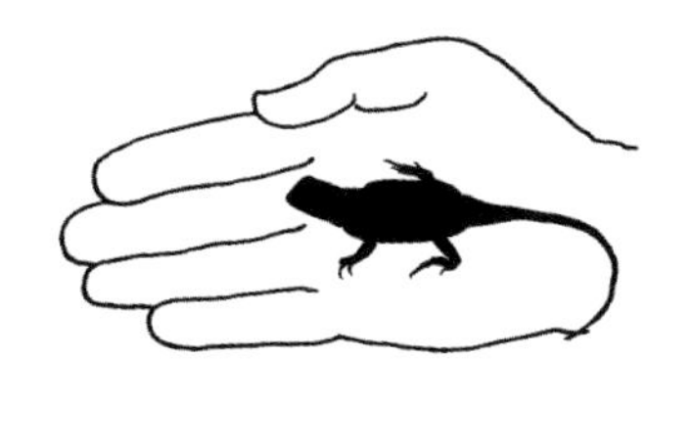

opposite page:
*Tympanocryptis pinguicolla* habitat. Little River area, Vic
photo P. Robertson

below:
*Tympanocryptis pinguicolla*. Breeding male. Little River, Vic. Photographed c 1967
photo F. and C. Collet

## GOLDFIELDS PEBBLE-MIMIC DRAGON

*Tympanocryptis pseudopsephos* Doughty, Kealley, Shoo & Melville, 2015

**DESCRIPTION:** SVL 56.5 mm. Rotund body, small bulbous head with blunt snout and protruding mouth, short neck and moderately short limbs and tail. Snout concave when viewed from side due to strongly protruding mouth. Scales on snout rugose with feeble keels and rostral scale 1–2 times as wide as high. Dorsal body scales include enlarged scales with raised spines in transverse rows of 4–7. Keels on scales of upper arm and dorsal surface of thigh are not aligned. A row of enlarged scales along the anterior dorsal edge of thigh forms a conspicuous ridge. Ventral scales smooth. Reddish-brown without lines but often with a dark charcoal "wash" over the head, limbs and back. Neck is dark brown, sometimes with three short white lines edged with black. Tail has about 8 pale bands edged anteriorly with black; the pale bands sometimes merging on posterior half of tail to form a pale stripe. Ventral surface pale, with yellow or light orange hue and sometimes dark stippling near gular fold. Femoral pores 0; preanal pores 2.

**KEY CHARACTERS:** Not known to overlap with any other *Tympanocrpytis*, Can be distinguished from the Pilbara *Tympanocryptis cephalus* species group in having enlarged dorsal scales in short transverse rows of 4–7 scales and having a concave snout. It differs further from *T. gigas* by a smaller body size, more rotund body shape, a conspicuous ridge on front of thigh formed by enlarged row of scales, and scales on top of thigh are homogeneous. It differs further from *T. diabolicus* and *T. fortescuensis* in having keels on upper arm and upper thigh scales not aligned (versus aligned) and smooth or slightly raised ventral scales (vs with low keels).

**DISTRIBUTION AND ECOLOGY:** The most widely distributed of the pebble-mimicking *T. cephalus* species-group. Distributed throughout the Goldfields region of WA, from near the south-eastern edge of the Pilbara, south to near Norseman and east to Neale Junction in the Great Victoria Desert. Occurs in Mallee and Mulga woodlands and sparsely vegetated shrublands, typically with stones strewn on open ground.

**BIOLOGY:** One of a small group of accomplished pebble-mimic dragons, with the head and body closely resembling two pebbles. When crouched immobile among scattered stones, it takes a keen eye to see through the disguise.

**NOTES:** WA's pebble-mimic earless dragons were reviewed by Doughty *et al* (2015). In previous publications, the name *T. cephalus* was applied to a composite of species occurring in stony arid areas of WA including throughout the range of *T. pseudosephos*. In virtually all illustrated publications, the images labelled as *T. cephalus* are actually the widely distributed *T. pseudosephos*.

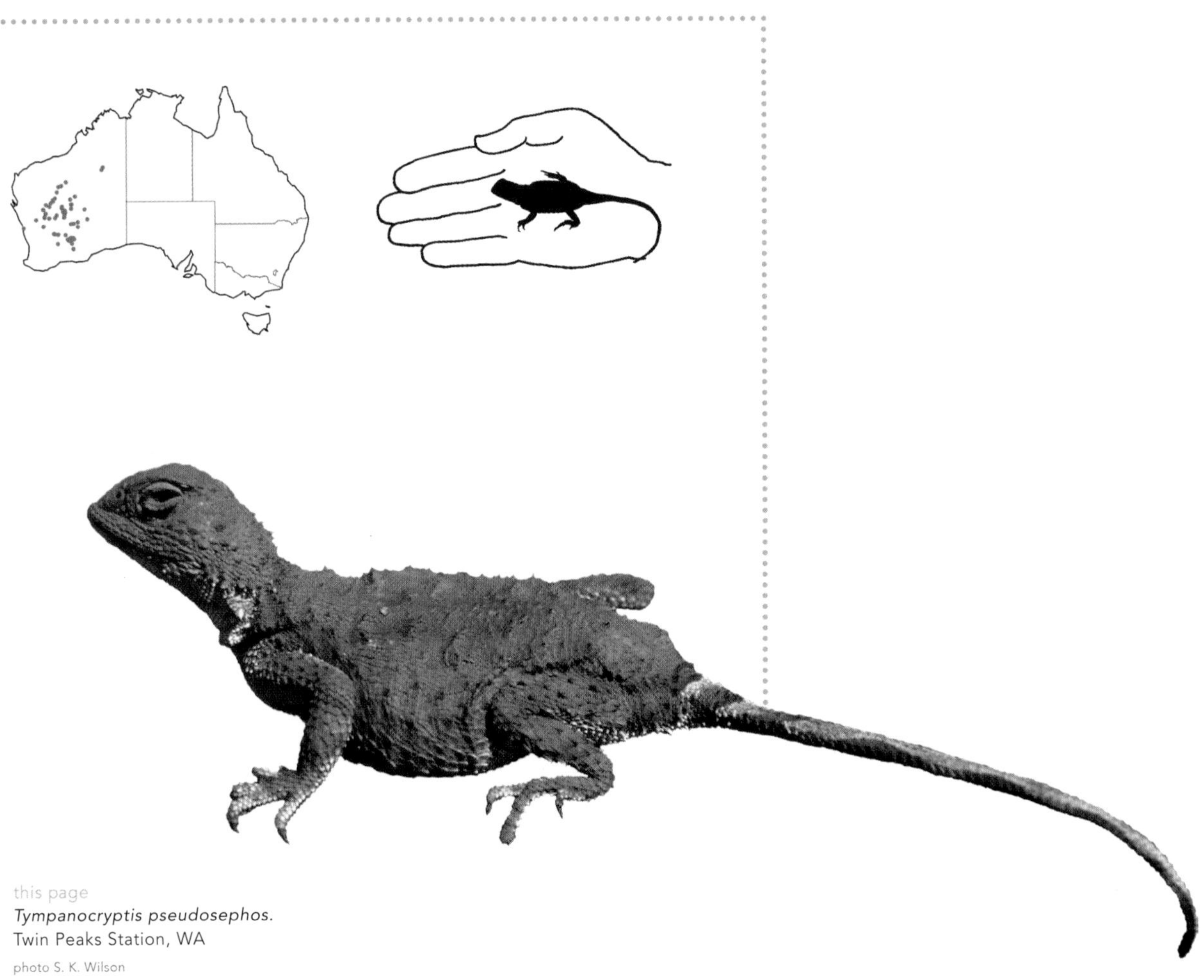

this page
*Tympanocryptis pseudosephos.*
Twin Peaks Station, WA
photo S. K. Wilson

opposite page
right:
*Tympanocryptis pseudosephos.*
Wilthorpe Station, WA
photo S. K. Wilson

left:
*Tympanocryptis pseudosephos*
habitat. Waldburg Station, WA
photo S. K. Wilson

# TYMPANOCRYPTIS

## EYREAN EARLESS DRAGON

*Tympanocryptis tetraporophora* Lucas & Frost, 1895

**DESCRIPTION:** SVL 54 mm. Relatively slender with weakly to strongly keeled scales on top of head and ventral surface, and small homogeneous dorsal body scales mixed with scattered enlarged mucronate scales. Body patterning variable. Reddish-brown or brown with a weak to strong pale vertebral stripe, and sometimes a pair of continuous or broken dorsolateral stripes overlying 4–6 darker brown broken bands. Narrow pale lateral stripe sometimes present. Some individuals have dark pigmentation on throat and upper chest. Lemon yellow on head and flanks has been observed in some individuals. Some females have pattern reduced to a broad grey band across back of head and neck. Femoral pores 2; preanal pores 2.

**KEY CHARACTERS:** Distribution overlaps with a number of other *Tympanocryptis*, including *T.* cf. *lineata*, *T. intima* and *T. pentalineata*. *T. tetraporophora* differs from the former two species by the presence of four pores and from *T. pentalineata* by usually lacking five longitudinal body stripes.

**DISTRIBUTION AND ECOLOGY:** Broadly distributed across arid and dry tropical central and eastern interior, ranging from the arid interior of SA, to semi-arid western NSW and into the tropical grasslands of the Gulf region in northern Qld and the Barkly Tablelands of the NT. Occupies a wide range of habitats, from stony desert plains, inland floodplains and black soil plains to tropical savannah grasslands.

**BIOLOGY:** Active during warmer months, although during the hottest months in interior activity is reduced to short periods during cooler times of day. Often observed perching on litter, earth clods, or small rocks along road verges. During hot weather it often stands erect on the hind limbs with the body angled into the sun. Known to use cracks in soil for shelter. In stony deserts up to eight individuals of both sexes have been found in burrow systems underneath small shrubs, presumably because this sparse vegetation is the only place they can burrow in this harsh environment. Average clutch size of 8 eggs, with a range of 1 to 14 recorded (Greer & Smith, 1999).

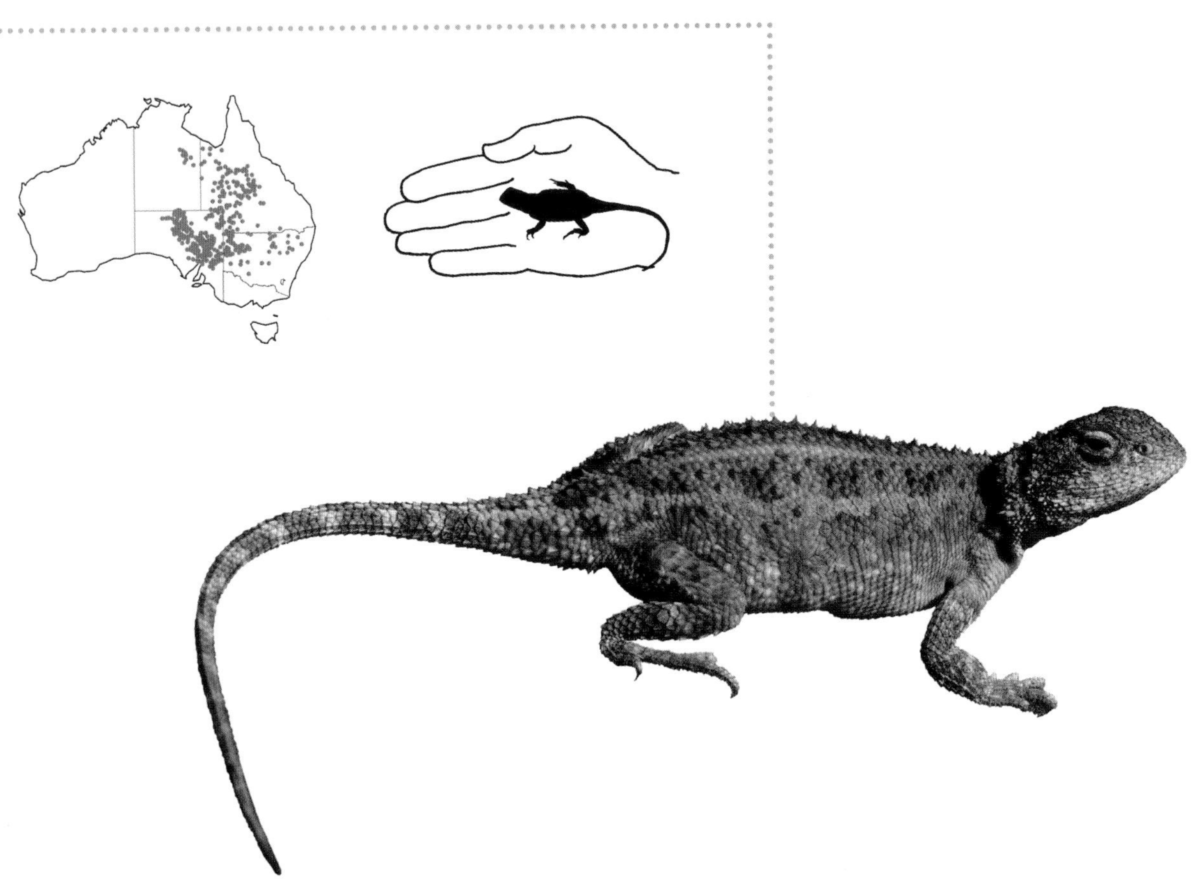

this page
*Tympanocryptis tetraporophora*. Gravid female.
Winton, Qld
photo S. K. Wilson

opposite page
right:
*Tympanocryptis tetraporophora*.
Brooklyn Station, Qld
photo S. K. Wilson

left:
*Tympanocryptis tetraporophora* habitat.
Winton area, Qld
photo S. K. Wilson

*Tympanocryptis tetraporophora.* Nilpena Station, SA
photo S. K. Wilson

## EVEN-SCALED EARLESS DRAGON

*Tympanocryptis uniformis* Mitchell, 1948

**DESCRIPTION:** SVL 55 mm. Small and stout with head almost as wide as long, a prominent canthal ridge, flat snout, short limbs and tail and a strong gular fold. Dorsal surface covered with keeled heterogenous scales but lacking enlarged tubercules. A few scattered enlarged tubercules on the neck. Ventral scales weakly keeled. Blue- grey with faint bands on the tail. Ventral surface dirty-white. Femoral pores 0; preanal pores 2.

**KEY CHARACTERS:** Does not overlap with any other *Tympanocryptis* species and differs from all other species in the uniform scales.

**DISTRIBUTION AND ECOLOGY:** Darwin area, Northern Territory.

**BIOLOGY:** Unknown.

**NOTES:** Only known with certainty from one specimen collected in 1911 near Darwin, NT. This is the only known *Tympanocryptis* specimen caught from this area, despite concerted searching over the decades. Previously it has been suggested that this species was also found around Wave Hill, NT, but recent genetic work has confirmed that this population is not *T. uniformis* but is *T. macra*. The specimen illustrated here is the preserved type specimen.

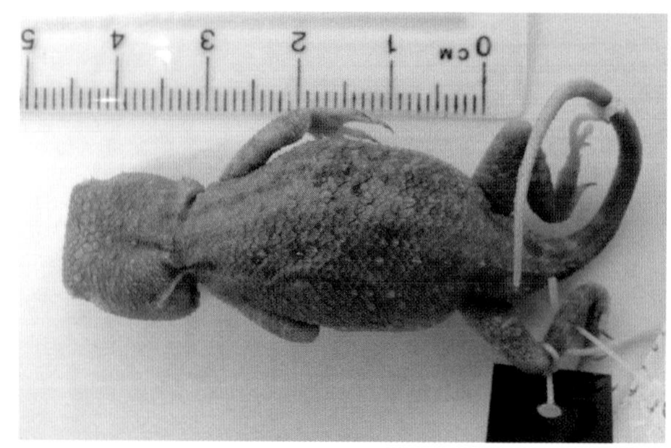

opposite page
right:
*Tympanocryptis uniformis.*
Preserved holotype. Darwin area, NT
photo M. Hutchinson

left:
*Tympanocryptis uniformis.*
Preserved holotype. Darwin area, NT
photo M. Hutchinson

## ROMA EARLESS DRAGON

*Tympanocryptis wilsoni* Melville, Smith, Hobson, Hunjan & Shoo, 2014

**DESCRIPTION:** SVL 47 mm. Dorsal scales are heterogeneous with un-keeled to strongly keeled scales and well-developed lateral and ventral body patterning. Fewer than 10 internasal scales. Brown to grey with prominent pale vertebral and dorsolateral stripes overlying a series of 4–5 broad dark bands. Usually little or no indication of a pale lateral stripe, but heavy brown-black speckling along flanks. Ventral pattern is concentrated on the head, throat and upper chest, extending back toward the lateral portions of the belly, with black-brown colouration more extensive than white interspaces. Three well defined pale spots on dorsal surface of snout: one above each nostril and one at end of snout. Some individuals have lemon yellow along their sides. Femoral pores 0–2; preanal pores 2.

**KEY CHARACTERS:** Similar to *T. tetraporaphora* but differs in having three well defined pale spots on dorsal surface of snout. Differs from *T. condaminensis* by ventral and lateral patterning with more black-brown than white colouration, the lateral stripe is usually absent (versus usually present), and often a pair of femoral pores (versus absent).

**DISTRIBUTION AND ECOLOGY:** Fragmented areas of native grasslands near Roma on the western Darling Downs, Qld: from Hodgson and Mt Abundance west to Amby area. Occurs in grasslands, dominated by Mitchell grasses. Observed along roadside verges bordering pasture and cropland and also along remaining stock-route reserves. A yellow flush recorded on the sides of the neck during March.

**BIOLOGY:** Little is known about this species. Most animals have been observed basking and perching on small rocks or earth mounds or along the edge of vegetation. When disturbed they flee down cracks in the soil or under rocks.

**NOTES:** Listed as Vulnerable in Qld and as Endangered internationally on the IUCN Red List (Melville *et al*, 2017c).

opposite page left:
*Tympanocryptis wilsoni* habitat.
Mt Abundance area near Roma
photo S. K. Wilson

opposite page right:
*Tympanocryptis wilsoni*.
Roma area, Qld
photo A. O'Grady

below:
*Tympanocryptis wilsoni*. Female.
Roma area, Qld
photo A. O'Grady

*Tympanocryptis wilsoni*. Mt Abundance area near Roma, Qld

photo S. K. Wilson

# REFERENCES


ALAND, K. 2008. Dragons, Family Agamidae. From Swan, M. (ed.) 2008. *Keeping and Breeding Australian Lizards*. Herp Books. Lilydale

AMEY, A. P., & WHITTIER, J. M. 2000. The annual reproductive cycle and sperm storage in the bearded dragon, *Pogona barbata*. *Australian Journal of Zoology*, 48, 411-419.

BRADSHAW, S. & MAIN, A. 1968. Behavioural attitudes and regulation of temperature in *Amphibolurus* lizards. *Journal of Zoology*, 154, 193-221.

BRADSHAW, S. D. 1986. *Ecophysiology of desert reptiles*, Academic Press.

BRAITHWAITE, R. W. 1987. Effects of fire regimes on lizards in the wet-dry tropics of Australia. *Journal of Tropical Ecology*, 3, 265-275.

BROOK, B. W. & GRIFFITHS, A. D. 2004. Fire management for the Frillneck Lizard *Chlamydosaurus kingii* in northern Australia. *Species Conservation and Management: Case Studies Using RAMAS GIS*, 312-325.

BROWN, D. 2012. *A guide to Australian dragons in captivity*. ABK Publications.

BYRNE, M., YEATES, D., JOSEPH, L., KEARNEY, M., BOWLER, J., WILLIAMS, M., COOPER, S., DONNELLAN, S., KEOGH, J. S. & LEYS, R. 2008. Birth of a biome: insights into the assembly and maintenance of the Australian arid zone biota. *Molecular Ecology*, 17, 4398-4417.

CADENA, V., RANKIN, K., SMITH, K. R., ENDLER, J. A., & STUART-FOX, D. 2017. Temperature-induced colour change varies seasonally in bearded dragon lizards. *Biological Journal of the Linnean Society*, 123, 422-430.

CATT, G., CRAIG, M. & SANDERSON, C. 2017. *Ctenophorus nguyarna. The IUCN Red List of Threatened Species 2017: eT83410198A83453738.*

CHAPMAN, A. & DELL, J. 1985. Biology and zoogeography of the amphibians and reptiles of the Western Australian wheatbelt. *Records of the WesternAustralian Museum*, 12, 1-46.

CHRISTIAN, K. A. & BEDFORD, G. S. 1995. Physiological consequences of filarial parasites in the frillneck lizard, *Chlamydosaurus kingii*, in northern Australia. *Canadian journal of zoology*, 73, 2302-2306.

CHRISTIAN, K. A., BEDFORD, G. S. & SHANNAHAN, S. T. 1996. Solar absorptance of some Australian lizards and its relationship to temperature. *Australian Journal of Zoology*, 44, 59-67.

COGGER, H. 1978. Reproductive cycles, fat body cycles and socio-sexual behaviour in the mallee dragon, *Amphibolurus fordi* (Lacertilia: Agamidae). *Australian Journal of Zoology*, 26, 653-672.

COOPER JR, W. E. & VITT, L. J. 2002. Distribution, extent, and evolution of plant consumption by lizards. *Journal of Zoology*, 257, 487-517.

DALY, B. G., DICKMAN, C. R. & CROWTHER, M. S. 2007. Selection of habitat components by two species of agamid lizards in sandridge desert, central Australia. *Austral Ecology*, 32, 825-833.

DALY, B. G., DICKMAN, C. R. & CROWTHER, M. S. 2008. Causes of habitat divergence in two species of agamid lizards in arid central Australia. *Ecology*, 89, 65-76.

DICKMAN, C. R., LETNIC, M. & MAHON, P. S. 1999. Population dynamics of two species of dragon lizards in arid Australia: the effects of rainfall. *Oecologia*, 119, 357-366.

DOUBE, B. 1975. The biology of the kangaroo tick, *Ornithodoros (Pavlovskyella) gurneyi* Warburton (Acarina: Argasidae), in the laboratory. *Journal of medical entomology*, 12, 240-243.

DOUGHTY, P., KEALLEY, L., SHOO, L. P. & MELVILLE, J. 2015. Revision of the Western Australian pebble-mimic dragon species-group (*Tympanocryptis cephalus*: Reptilia: Agamidae). *Zootaxa*, 4039, 85-117.

DOUGHTY, P., MARYAN, B., MELVILLE, J. & AUSTIN, J. 2007. A new species of *Ctenophorus* (Lacertilia: agamidae) from lake disappointment, Western Australia. *Herpetologica*, 63, 72-86.

DOUGHTY, P., KEALLEY, L. & MELVILLE, J. 2012. Taxonomic assessment of *Diporiphora* (Reptilia: Agamidae) dragon lizards from the western arid zone of Australia. Zootaxa 3518, 1-24.

DOUGHTY, P., TEALE, R., MELVILLE, J., LLOYD, R., GAIKHORST, G., CRAIG, M. & SANDERSON, C. 2017. *Diporiphora vescus*. The IUCN Red List of Threatened Species 2017: e.T83491786A83491797.

DRISCOLL, D. A. 2004. Extinction and outbreaks accompany fragmentation of a reptile community. *Ecological Applications*, 14, 220-240.

EDWARDS, D. L., & MELVILLE, J. 2010. Phylogeographic analysis detects congruent biogeographic patterns between a woodland agamid and Australian wet tropics taxa despite disparate evolutionary trajectories. *Journal of Biogeography*, 37, 1543-1556.

EDWARDS, D. L., MELVILLE, J., JOSEPH, L., KEOGH, J. S., STEPPAN, S. J. & BRONSTEIN, J. L. 2015. Ecological divergence, adaptive diversification, and the evolution of social signaling traits: an empirical study in Arid Australian Lizards. *The American Naturalist*, 186, E144-E161.

EHMANN, H. & STRAHAN, R. 1992. *Encyclopedia of Australian animals: reptiles*. Angus & Robertson.

EVANS, S. E. & JONES, M. E. 2010. The origin, early history and diversification of lepidosauromorph reptiles. *New aspects of Mesozoic biodiversity*. Springer.

EZAZ, T., QUINN, A. E., SARRE, S. D., O'MEALLY, D., GEORGES, A. & GRAVES, J. A. M. 2009. Molecular marker suggests rapid changes of sex-determining mechanisms in Australian dragon lizards. *Chromosome Research*, 17, 91-98.

FITZGERALD, M., SHINE, R. & LEMCKERT, F. 2004. Life history attributes of the threatened Australian snake (Stephen's banded snake *Hoplocephalus stephensii*, Elapidae). *Biological Conservation*, 119, 121-128.

FRANK, A. S., DICKMAN, C. R., WARDLE, G. M. & GREENVILLE, A. C. 2013. Interactions of grazing history, cattle removal and time since rain drive divergent short-term responses by desert biota. *PLoS One*, 8, e68466.

FRY, B. G., VIDAL, N., NORMAN, J. A., VONK, F. J., SCHEIB, H., RAMJAN, S. R., KURUPPU, S., FUNG, K., HEDGES, S. B. & RICHARDSON, M. K. 2006. Early evolution of the venom system in lizards and snakes. *Nature*, 439, 584-588.

GARCIA, J. E., ROHR, D. & DYER, A. G. 2013. Trade-off between camouflage and sexual dimorphism revealed by UV digital imaging: the case of Australian Mallee dragons (*Ctenophorus fordi*). *Journal of Experimental Biology*, 216, 4290-4298.

GEORGES, A., LI, Q., LIAN, J., O'MEALLY, D., DEAKIN, J., WANG, Z., ZHANG, P., FUJITA, M., PATEL, H. R. & HOLLELEY, C. E. 2015. High-coverage sequencing and annotated assembly of the genome of the Australian dragon lizard *Pogona vitticeps*. *Gigascience*, 4, 45.

GIBBONS, J. R. 1979. The hind leg pushup display of the *Amphibolurus decresii* species complex (Lacertilia: Agamidae). *Copeia*, 29-40.

GLEN, A. & DICKMAN, C. 2006. Diet of the spotted-tailed quoll (*Dasyurus maculatus*) in eastern Australia: effects of season, sex and size. *Journal of Zoology*, 269, 241-248.

GREER, A. & SMITH, S. 1999. Aspects of the morphology and reproductive biology of the Australian earless dragon lizard *Tympanocryptis tetraporophora*. *Australian Zoologist*, 31, 55-70.

GREER, A. E. 1989. *The biology and evolution of Australian lizards*, Surrey Beatty and Sons.

GRIFFITHS, A. D. 1999. Demography and home range of the frillneck lizard, *Chlamydosaurus kingii* (Agamidae), in northern Australia. *Copeia*, 1089-1096.

HAMILTON, D. G., WHITING, M. J. & PRYKE, S. R. 2013. Fiery frills: carotenoid-based coloration predicts contest success in frillneck lizards. *Behavioral Ecology*, 24, 1138-1149.

HARLOW, P. S. 2004. Temperature-dependent sex determination in lizards. *Temperature-dependent sex determination in vertebrates*, 42-52.

HARLOW, P. S. & PRICE, A. 2000. Incubation temperature determines hatchling sex in Australian rock dragons (Agamidae: Genus *Ctenophorus*). *Copeia*, 2000, 958-964.

HARLOW, P. S. & SHINE, R. 1999. Temperature-dependent sex determination in the frillneck lizard, Chlamydosaurus kingii (Agamidae). Herpetologica, 205-212.

HARLOW, P. S. & TAYLOR, J. E. 2000. Reproductive ecology of the jacky dragon (*Amphibolurus muricatus*): an agamid lizard with temperature-dependent sex determination. *Austral Ecology*, 25, 640-652.

HOCKNULL, S. A., ZHAO, J.-X., FENG, Y.-X. & WEBB, G. E. 2007. Responses of Quaternary rainforest vertebrates to climate change in Australia. *Earth and Planetary Science Letters*, 264, 317-331.

HOLLELEY, C. E., O'MEALLY, D., SARRE, S. D., GRAVES, J. A. M., EZAZ, T., MATSUBARA, K., AZAD, B., ZHANG, X. & GEORGES, A. 2015. Sex reversal triggers the rapid transition from genetic to temperature-dependent sex. *Nature*, 523, 79-82.

HOUSTON, T. F. & HUTCHINSON, M. 1998, *Dragon lizards and goannas of South Australia, 2nd ed*, South Australian Museum, Adelaide, SA

HUGALL, A. F., FOSTER, R., HUTCHINSON, M. & LEE, M. S. 2008. Phylogeny of Australasian agamid lizards based on nuclear and mitochondrial genes: implications for morphological evolution and biogeography. *Biological Journal of the Linnean Society*, 93, 343-358.

HUSBAND, G. 1979. Herpetological notes. Range extension for *Chelosania brunnea*. *Herpetofauna*, 10, 29.

HUTCHINSON, M. N. & HUTCHINSON, R. G. 2011. Karyotypes of Moloch and *Chelosania* (Squamata: Acrodonta). *Journal of Herpetology*, 45, 216-218.

JAMES, C. & SHINE, R. 1988. Life-history strateqies of Australian lizards: a comparison between the tropics and the temperate zone. *Oecologia*, 75, 307-316.

JESSOP, T. S., CHAN, R. & STUART-FOX, D. 2009. Sex steroid correlates of female-specific colouration, behaviour and reproductive state in Lake Eyre dragon lizards, *Ctenophorus maculosus*. *Journal of Comparative Physiology A*, 195, 619-630.

JOHNSTON, G. 1999. Reproductive biology of the Peninsula dragon lizard, *Ctenophorus fionni*. *Journal of Herpetology*, 33, 694-698.

JOHNSTON, G. 2011. Growth and survivorship as proximate causes of sexual size dimorphism in peninsula dragon lizards *Ctenophorus fionni*. *Austral Ecology*, 36, 117-125.

KAVANAGH, R. P. & STANTON, M. A. 2005. Vertebrate species assemblages and species sensitivity to logging in the forests of north-eastern New South Wales. *Forest Ecology and Management*, 209, 309-341.

KENT, D. 1987. Notes on the biology and osteology of *Amphibolurus diemensis* (Gray, 1841), the Mountain Dragon. *Victorian Naturalist*, 104, 101-104.

KUTT, A., BATEMAN, B. & VANDERDUYS, E. 2011. Lizard diversity on a rainforest–savanna altitude gradient in north-eastern Australia. *Australian Journal of Zoology*, 59, 86-94.

KUTT, A. S. 2011. The diet of the feral cat (*Felis catus*) in north-eastern Australia. *Acta Theriologica*, 56, 157-169.

LEBAS, N. 2001. Microsatellite determination of male reproductive success in a natural population of the territorial ornate dragon lizard, *Ctenophorus ornatus*. *Molecular Ecology*, 10, 193-203.

LEBAS, N. R. & MARSHALL, N. J. 2000. The role of colour in signalling and male choice in the agamid lizard *Ctenophorus ornatus*. *Proceedings of the Royal Society of London B: Biological Sciences*, 267, 445-452.

LETNIC, M., DICKMAN, C., TISCHLER, M., TAMAYO, B. & BEH, C.-L. 2004. The responses of small mammals and lizards to post-fire succession and rainfall in arid Australia. *Journal of arid environments*, 59, 85-114.

LEVY, E., KENNINGTON, J. W., TOMKINS, J. L. & LEBAS, N. R. 2010. Land clearing reduces gene flow in the granite outcrop-dwelling lizard, *Ctenophorus ornatus*. *Molecular ecology*, 19, 4192-4203.

LEVY, E., KENNINGTON, W. J., TOMKINS, J. L. & LEBAS, N. R. 2012. Phylogeography and population genetic structure of the ornate dragon lizard, *Ctenophorus ornatus*. *PloS one*, 7, e46351.

LITTLEFORD-COLQUHOUN, B. L., CLEMENTE, C., WHITING, M. J., ORTIZ-BARRIENTOS, D. & FRÈRE, C. H. 2017. Archipelagos of the Anthropocene: rapid and extensive differentiation of native terrestrial vertebrates in a single metropolis. *Molecular ecology*, 26, 2466-2481.

LOSOS, J. B. 1987. Postures of the military dragon (*Ctenophorus isolepis*) in relation to substrate temperature. *Amphibia-Reptilia*, 8, 419-423.

MACKERRAS, M. J. 1961. The haematozoa of Australian reptiles. *Australian Journal of Zoology*, 9, 61-122.

MAYES, P., THOMPSON, G. & WITHERS, P. 2005. Diet and foraging behaviour of the semi-aquatic *Varanus mertensi* (Reptilia: Varanidae). *Wildlife Research*, 32, 67-74.

MCKAY, J. L. 2011. *Frillneck!*, Darwin, Valvolandia Publishing.

MCLEAN, C. A., LUTZ, A., RANKIN, K. J., STUART-FOX, D., & MOUSSALLI, A. 2017. Revealing the biochemical and genetic basis of color variation in a polymorphic lizard. *Molecular biology and evolution*, 34, 1924-1935.

MCLEAN, C. A., MOUSSALLI, A. & STUART-FOX, D. 2010. The predation cost of female resistance. *Behavioral Ecology*, 21, 861-867.

MCLEAN, C. A. & STUART-FOX, D. 2015. Rival assessment and comparison of morphological and performance-based predictors of fighting ability in Lake Eyre dragon lizards, *Ctenophorus maculosus*. *Behavioral Ecology and Sociobiology*, 69, 523-531.

MCLEAN, C. A., STUART-FOX, D. & MOUSSALLI, A. 2015. Environment, but not genetic divergence, influences geographic variation in colour morph frequencies in a lizard. *BMC evolutionary biology*, 15, 156.

MELVILLE, J., CHAPLIN, K., HUTCHINSON, M., SUMNER, J., GRUBER, B., MACDONALD, A.J. & SARRE, S.D. 2019a. Taxonomy and conservation of grassland earless dragons: new species and an assessment of the first possible extinction of a reptile on mainland Australia. Royal Society Open Science, p.190233.

MELVILLE, J., HUTCHINSON, M., WILSON, S., HOBSON, R. & VENZ, M 2017a. *Tympanocryptis condaminensis. The IUCN Red List of Threatened Species 2017: e.T83494956A83494963.*

MELVILLE, J., HUTCHINSON, M., ROBERTSON, P. & MICHAEL, D. 2017b. *Ctenophorus mirrityana. The IUCN Red List of Threatened Species 2017: e.T83410195A83453733.*

MELVILLE, J., HUTCHINSON, M., CLEMANN, N., ROBERTSON, P. & MICHAEL, D. 2018. Tympanocryptis pinguicolla. The IUCN Red List of Threatened Species 2018: e.T22579A83494675.

MELVILLE, J., WILSON, S. & HOBSON, R. 2017c. *Tympanocryptis wilsoni. The IUCN Red List of Threatened Species 2017: e T83495150A83495160.*

MELVILLE, J., GOEBEL, S., STARR, C., KEOGH, J. S. & AUSTIN, J. J. 2007. Conservation genetics and species status of an endangered Australian dragon, *Tympanocryptis pinguicolla* (Reptilia: Agamidae). *Conservation Genetics*, 8, 185-195.

MELVILLE, J., HAINES, M. L., BOYSEN, K., HODKINSON, L., KILIAN, A., DATE, K. L. S., POTVIN, D. A. & PARRIS, K. M. 2017d. Identifying hybridization and admixture using SNPs: application of the DArTseq platform in phylogeographic research on vertebrates. *Royal Society open science*, 4, 161061.

MELVILLE, J., HAINES, M. L., HALE, J., CHAPPLE, S. & RITCHIE, E. G. 2016. Concordance in phylogeography and ecological niche modelling identify dispersal corridors for reptiles in arid Australia. *Journal of Biogeography*, 43, 1844-1855.

MELVILLE, J., HARMON, L. J. & LOSOS, J. B. 2006. Intercontinental community convergence of ecology and morphology in desert lizards. *Proceedings of the Royal Society of London B: Biological Sciences*, 273, 557-563.

MELVILLE, J., RITCHIE, E., CHAPPLE, S., GLOR, R. & SCHULTE, J. 2011. Evolutionary origins and diversification of dragon lizards in Australia's tropical savannas. *Molecular Phylogenetics and Evolution*, 58, 257-270.

MELVILLE, J., RITCHIE, E., CHAPPLE, S., GLOR, R. & SCHULTE, J. 2018. Diversity in Australia's tropical savannas: An integrative taxonomic revision of agamid lizards from the genera *Amphibolurus* and *Lophognathus* (Lacertilia: Agamidae). *Memoirs of Museum Victoria* 77: 41–61

MELVILLE, J., & SCHULTE, II, J. A. 2001. Correlates of active body temperatures and microhabitat occupation in nine species of central Australian agamid lizards. *Austral Ecology*, 26, 660-669.

MELVILLE, J., SCHULTE, J.A. & LARSON, A. 2001. A molecular phylogenetic study of ecological diversification in the Australian lizard genus *Ctenophorus*. Journal of Experimental Zoology 291, 339-353.

MELVILLE, J., SMITH, K., HOBSON, R., HUNJAN, S. & SHOO, L. 2014. The role of integrative taxonomy in the conservation management of cryptic species: the taxonomic status of endangered earless dragons (Agamidae: *Tympanocryptis*) in the grasslands of Queensland, Australia. *PloS one*, 9, e101847.

MELVILLE, J., SMITH DATE, K., HORNER, P., & DOUGHTY, P. 2019. Taxonomic revision of dragon lizards in the genus *Diporiphora* (Reptilia: Agamidae) from the Australian monsoon tropics. *Memoirs of Museum Victoria*. 78: 23–55

MEYERS, J. J. & HERREL, A. 2005. Prey capture kinematics of ant-eating lizards. Journal of Experimental Biology, 208, 113-127.

MITCHELL, F. 1973. Studies on the ecology of the agamid lizard Amphibolurus maculosus (Mitchell). Transactions of the Royal Society of South Australia, 97, 47-76.

NG, J., CLEMANN, N., CHAPPLE, S. N. & MELVILLE, J. 2014. Phylogeographic evidence links the threatened 'Grampians' Mountain Dragon (Rankinia diemensis Grampians) with Tasmanian populations: conservation implications in south-eastern Australia. Conservation genetics, 15, 363-373.

NIEJALKE, D. P. 2006. Reproduction by a small agamid lizard, Ctenophorus pictus, during contrasting seasons. Herpetologica, 62, 409-420.

OLIVER, P. M. & HUGALL, A. F. 2017. Phylogenetic evidence for mid-Cenozoiç turnover of a diverse continental biota. Nature ecology & evolution, 1, 1896.

OLSSON, M. 1995. Forced copulation and costly female resistance behavior in the Lake Eyre dragon, Ctenophorus maculosus. Herpetologica, 19-24.

OLSSON, M. 2001a. No female mate choice in Mallee dragon lizards, Ctenophorus fordi. Evolutionary Ecology, 15, 129-141.

OLSSON, M. 2001b. 'Voyeurism'prolongs copulation in the dragon lizard Ctenophorus fordi. Behavioral Ecology and Sociobiology, 50, 378-381.

OLSSON, M., HEALEY, M., PERRIN, C., WILSON, M. & TOBLER, M. 2012a. Sex-specific SOD levels and DNA damage in painted dragon lizards (Ctenophorus pictus). Oecologia, 170, 917-924.

OLSSON, M., HEALEY, M., WILSON, M. & TOBLER, M. 2012b. Polymorphic male color morphs visualized with steroids in monomorphic females: a tool for designing analysis of sex-limited trait inheritance. Journal of Experimental Biology, 215, 575-577.

OLSSON, M., SCHWARTZ, T., ULLER, T., & HEALEY, M. 2009. Effects of sperm storage and male colour on probability of paternity in a polychromatic lizard. Animal Behaviour, 77, 419-424.

OLSSON, M., STUART-FOX, D. & BALLEN, C. 2013. Genetics and e volution of colour patterns in reptiles. Seminars in cell & developmental biology, Elsevier, 529-541.

ORD, T. & STUART-FOX, D. 2006. Ornament evolution in dragon lizards: multiple gains and widespread losses reveal a complex history of evolutionary change. Journal of Evolutionary Biology, 19, 797-808.

PALMER, R., PEARSON, D. J., COWAN, M. A. & DOUGHTY, P. 2013. Islands and scales: a biogeographic survey of reptiles on Kimberley islands, Western Australia. Records of the Western Australian Museum, 81, 183-204.

PASTRO, L. A., DICKMAN, C. R. & LETNIC, M. 2013. Effects of wildfire, rainfall and region on desert lizard assemblages: the importance of multi-scale processes. Oecologia, 173, 603-614.

PETERS, R. A. & ORD, T. J. 2003. Display response of the Jacky Dragon, Amphibolurus muricatus (Lacertilia: Agamidae), to intruders: A semi-Markovian process. Austral Ecology, 28, 499-506.

PETERSON, M., SHEA, G., JOHNSTON, G. & MILLER, B. 1994. Notes on the morphology and biology of Ctenophorus mckenziei (Storr, 1981) (Squamata: Agamidae). Transactions of the Royal Society of South Australia, 118, 237-244.

PIANKA, E. R. 1971a. Ecology of the agamid lizard Amphibolurus isolepis in Western Australia. Copeia, 527-536.

PIANKA, E. 1971b. Notes on the biology of Amphibolurus cristatus and Amphibolurus scutulatus. West. Aust. Nat, 12, 36-41.

PIANKA, E. R. 2013a. Notes on the ecology and natural history of two uncommon terrestrial agamid lizards Ctenophorus clayi and C. fordi in the Great Victoria desert of Western Australia. Western Australian Naturalist, 29, 85-93.

PIANKA, E. R. 2013b. Notes on the natural history of the rarely recorded agamid lizard Caimanops amphiboluroides in Western Australia. Western Australian Naturalist, 29, 99-102.

PIANKA, E. R. 2014. Notes on the ecology and natural history of Ctenophorus caudicinctus (Agamidae) in Western Australia. Western Australian Naturalist 30, 226-230.

PIANKA, G., PIANKA, E. & THOMPSON, G. 1996. Egg laying by thorny devils (Moloch horridus) under natural conditions. Journal of the Royal Society of Western Australia, 79, 195-197.

PIANKA, G., PIANKA, E. & THOMPSON, G. 1998. Natural history of thorny devils Moloch horridus (Lacertilia: Agamidae) in the Great Victoria Desert. Journal of the Royal Society of Western Australia, 81, 183-190.

POWNEY, G., GRENYER, R., ORME, C., OWENS, I. & MEIRI, S. 2010. Hot, dry and different: Australian lizard richness is unlike that of mammals, amphibians and birds. Global Ecology and Biogeography, 19, 386-396.

PYRON, R. A., BURBRINK, F. T. & WIENS, J. J. 2013. A phylogeny and revised classification of Squamata, including 4161 species of lizards and snakes. BMC evolutionary biology, 13, 93.

RADWAN, J., KUDUK, K., LEVY, E., LEBAS, N. & BABIK, W. 2014. Parasite load and MHC diversity in undisturbed and agriculturally modified habitats of the ornate dragon lizard. Molecular ecology, 23, 5966-5978.

RICE, G. & BRADSHAW, S. 1980. Changes in dermal reflectance and vascularity and their effects on thermoregulation in Amphibolurus nuchalis (reptilia: Agamidae). Journal of Comparative Physiology B: Biochemical, Systemic, and Environmental Physiology, 135, 139-146.

ROBERTSON, P., SILVEIRA, C., COVENTRY, J., EYLES, D., WILLIAMS, A. & SLUITER, I. 2005. Examination of the effects of the 2002 wildfire in the Big Desert on terrestrial vertebrates. Interim report.

RUMMERY, C., SHINE, R., HOUSTON, D. L. & THOMPSON, M. B. 1995. Thermal biology of the Australian forest dragon, Hypsilurus spinipes (Agamidae). Copeia, 818-827.

SADLIER, R. 1990. The Terrestrial and Semiaquatic Reptiles (Lacertilia, Serpentes) of the Magela Creek Region, Northern Territory, Australian Government Pub. Service.

SADLIER, R. & PRESSEY, R. 1994. Reptiles and amphibians of particular conservation concern in the western division of New South Wales: a preliminary review. Biological Conservation, 69, 41-54.

SCHÄUBLE, C. S., & GRIGG, G. C. 1998. Thermal ecology of the Australian agamid Pogona barbata. Oecologia, 114, 461-470.

SCHULTE, II, J. A., MELVILLE, J. & LARSON, A. 2003. Molecular phylogenetic evidence for ancient divergence of lizard taxa on either side of Wallace's Line. Proceedings of the Royal Society of London B: Biological Sciences, 270, 597-603.

SHERBROOKE, W. C., SCARDINO, A. J., DE NYS, R. & SCHWARZKOPF, L. 2007. Functional morphology of scale hinges used to transport water: convergent drinking adaptations in desert lizards (Moloch horridus and Phrynosoma cornutum). Zoomorphology, 126, 89-102.

SHINE, R. 1990. Function and evolution of the frill of the frillneck lizard, Chlamydosaurus kingii (Sauria: Agamidae). Biological Journal of the Linnean Society, 40, 11-20.

SHINE, R. & LAMBECK, R. 1989. Ecology of Frillneck Lizards, Chlamydosaurus kingii (Agamidae), in Tropical Australia. Wildlife Research, 16, 491-500

SHOO, L. P., ROSE, R., DOUGHTY, P., AUSTIN, J. J., & MELVILLE, J. 2008. Diversification patterns of pebble-mimic dragons are consistent with historical disruption of important habitat corridors in arid Australia. Molecular Phylogenetics and Evolution, 48, 528-542.

SMITH, A. L., BULL, C. M. & DRISCOLL, D. A. 2013. Skeletochronological analysis of age in three 'fire-specialist' lizard species. South Aust. Nat, 87, 6-17.

SMITH, J. & SCHWANER, T. 1981. Notes on Reproduction by Captive Amphibolurus Nullarbor (Sauria: Agamidae). Transactions of the Royal Society of South Australia. Adelaide, 105, 215-216.

SMITH, K. L., HARMON, L. J., SHOO, L. P. & MELVILLE, J. 2011. Evidence of constrained phenotypic evolution in a cryptic species complex of agamid lizards. Evolution, 65, 976-992.

SMITH, K. R., CADENA, V., ENDLER, J. A., KEARNEY, M. R., PORTER, W. P. & STUART-FOX, D. 2016b. Color change for thermoregulation versus camouflage in free-ranging lizards. The American Naturalist, 188, 668-678.

SMITH, K. R., CADENA, V., ENDLER, J. A., PORTER, W. P., KEARNEY, M. R. & STUART-FOX, D. 2016a. Colour change on different body regions provides thermal and signalling advantages in bearded dragon lizards. Proc. R. Soc. B, 283, 20160626.

SOKOLOVA, Y. Y., SAKAGUCHI, K. & PAULSEN, D. B. 2016. Establishing a new species Encephalitozoon pogonae for the microsporidian parasite of inland bearded dragon Pogona vitticeps Ahl 1927 (Reptilia, Squamata, Agamidae). Journal of Eukaryotic Microbiology, 63, 524-535.

STARR, C. R. & LEUNG, L. K.-P. 2006. Habitat use by the Darling Downs population of the grassland earless dragon: implications for conservation. Journal of Wildlife Management, 70, 897-903.

STEVENS, T. A., EVANS, M. C., OSBORNE, W. S. & SARRE, S. D. 2010. Home ranges of, and habitat use by, the grassland earless dragon (Tympanocryptis pinguicolla) in remnant native grasslands near Canberra. Australian Journal of Zoology, 58, 76-84.

STORR, G. 1967. Geographic races of the agamid lizard Amphibolurus caudicinctus. Journal of the Royal Society of Western Australia, 50, 49-56.

STORR, G. 1974. Agamid lizards of the genera Caimanops, Physignathus and Diporiphora in Western Australia and Northern Territory. Records of the Western Australian Museum, 3, 121-146.

STORR, G. & HANLON, T. 1980. Herpetofauna of the Exmouth region, Western Australia. Records of the Western Australian Museum, 8, 423-439.

STRICKLAND, K., PATTERSON, E.M. & FRÈRE, C.H. 2018. Eastern water dragons use alternative social tactics at different local densities. Behavioral Ecology and Sociobiology 72, 148.

STUART-FOX, D. M., MOUSSALLI, A., JOHNSTON, G. R. & OWENS, I. P. 2004. Evolution of color variation in dragon lizards: quantitative tests of the role of crypsis and local adaptation. Evolution, 58, 1549-1559.

STUART-FOX, D. M., MOUSSALLI, A., MARSHALL, N. J. & OWENS, I. P. 2003. Conspicuous males suffer higher predation risk: visual modelling and experimental evidence from lizards. Animal Behaviour, 66, 541-550.

STUART-SMITH, J., SWAIN, R. & WELLING, A. 2005. Reproductive ecology of the mountain dragon, Rankinia (Tympanocryptis) diemensis (Reptilia: Squamata: Agamidae) in Tasmania. Papers and Proceedings of the Royal Society of Tasmania, 2005. 23-28.

STUART-SMITH, J. F., STUART-SMITH, R. D., SWAIN, R. & WAPSTRA, E. 2008. Size dimorphism in Rankinia [Tympanocryptis] diemensis (Family Agamidae): sex-specific patterns and geographic variation. Biological Journal of the Linnean Society, 94, 699-709.

TATTERSALL, G. J. & GERLACH, R. M. 2005. Hypoxia progressively lowers thermal gaping thresholds in bearded dragons, Pogona vitticeps. Journal of experimental biology, 208, 3321-3330.

THOMPSON, G. & THOMPSON, S. 2001. Behaviour and spatial ecology of Gilbert's dragon Lophognathus gilberti (Agamidae: Reptilia). Journal of the Royal Society of Western Australia, 84, 153-158.

THOMPSON, G. G. & WITHERS, P. C. 2005. The relationship between size-free body shape and choice of retreat for Western Australian Ctenophorus (Agamidae) dragon lizards. Amphibia-Reptilia, 26, 65-72.

THOMPSON, S., THOMPSON, G. & OATES, J. 2008. Range extension of the Western Heath Dragon, Rankinia adelaidensis adelaidensis (Squamata: Agamidae). Journal of the Royal Society of Western Australia, 91, 207-208.

TRAINOR, C. & WOINARSKI, J. 1994. Responses of lizards to three experiments fires in the savanna forests of Kakadu National Park. Wildlife Research, 21, 131-147.

TRAINOR, C. R. 2005. Distribution and natural history of the cryptic Chameleon Dragon Chelosania brunnea: a review of records. Northern Territory Naturalist, 34.

TREMBATH, D. F., FEARN, S. & UNDHEIM, E. A. B. 2009. Natural history of the slaty grey snake (Stegonotus cucullatus) (Serpentes: Colubridae) from tropical north Queensland, Australia. Australian Journal of Zoology, 57, 119-124.

UJVARI, B., DOWTON, M. & MADSEN, T. 2008. Population genetic structure, gene flow and sex-biased dispersal in frillneck lizards (Chlamydosaurus kingii). Molecular Ecology, 17, 3557-3564.

ULLER, T., SCHWARTZ, T., KOGLIN, T. & OLSSON, M. 2013. Sperm storage and sperm competition across ovarian cycles in the dragon lizard, Ctenophorus fordi. Journal of Experimental Zoology Part A: Ecological Genetics and Physiology, 319, 404-408.

WARNER, D. A. & SHINE, R. 2008. Determinants of Dispersal Distance in Free-Ranging Juvenile Lizards. Ethology, 114, 361-368.

WARNER, D. A., ULLER, T. & SHINE, R. 2009. Fitness effects of the timing of hatching may drive the evolution of temperature-dependent sex determination in short-lived lizards. Evolutionary ecology, 23, 281.

WEIGEL, J. 1989. Maintenance and breeding of the Superb dragon Diporiphora superba at the Australian Reptile Park, Gosford. International Zoo Yearbook, 28, 122-126.

WILLIAMS, J. R., DRISCOLL, D. A. & BULL, C. M. 2012. Roadside connectivity does not increase reptile abundance or richness in a fragmented mallee landscape. Austral Ecology, 37, 383-391.

WILSON, S., 2012. Australian lizards: a natural history. CSIRO Publishing.

WILSON, S. K. & KNOWLES, D. G. 1988. Australia's reptiles: a photographic reference to the terrestrial reptiles of Australia, Collins Australia.

WITHERS, P. & DICKMAN, C. 1995. The role of diet in determining water, energy and salt intake in the thorny devil Moloch horridus (Lacertilia: Agamidae). Journal of the Royal Society of Western Australia, 78, 3.

WOTHERSPOON, D. & BURGIN, S. 2011. Allometric variation among juvenile, adult male and female eastern bearded dragons Pogona barbata (Cuvier, 1829), with comments on the behavioural implications. Zoology, 114, 23-28.

WOTHERSPOON, D. & BURGIN, S. 2016. Sex and ontogenetic dietary shift in Pogona barbata, the Australian eastern bearded dragon. Australian Journal of Zoology, 64, 14-20.

ZHENG, Y. & WIENS, J. J. 2016. Combining phylogenomic and supermatrix approaches, and a time-calibrated phylogeny for squamate reptiles (lizards and snakes) based on 52 genes and 4162 species. Molecular phylogenetics and evolution, 94, 537-547.

# SPECIES INDEX

## SCIENTIFIC NAMES

# SPECIES INDEX

## COMMON NAMES

# ACKNOWLEDGEMENTS

It has taken much more than two co-authors to create this book. Photographers have generously provided their images, fellow herpetologists have made their observations available to us and offered useful comments, and we have shared valuable field time with friends and colleagues. We have also had vital and much appreciated support on the home front.

To the photographers whose images appear in this book, we are aware that the time, effort and expense involved in finding and photographing elusive species are not always apparent in the finished product. For the use of your excellent photographs, we would like to thank: Mark Binns, Kirilee Chaplin, Rob Browne-Cooper, Nick Clemann, Ross Coupland, Frank and Chris Collet, Melissa Bruton, Brian Bush, Matt Clancy, Pat Cullen, Jordan de Jong, Adam Elliott, Angus Emmott, Aaron Fenner, Stewart Ford, Glen Gaikhorst, Rich Glor, David Goldney, Greg Harold, Paul Horner, Mark Hutchinson, Dave Knowles, Rosie Koch, Ray Lloyd, Stewart Macdonald, Stephen Mahony, Brad Maryan, Andrew O'Grady, Paul Oliver, David Paul, Mark Pestov, Peter Robertson, Akash Samuel, Brendan Schembri, Shawn Scott, Emma Sherratt, Garry Stephenson, Devi Stuart-Fox, Gerry Swan, Eric Vanderduys, Jordan Vos, Philipp Wagner, Gavin Waters, Justin Wright and Stephen Zozaya.

For valuable time spent in the field, and for the many conversations about our shared interests over the years, we acknowledge: Nick Clemann, Garth Coupland, Paul Doughty, Angus Emmott, Mark Hanlon, Paul Horner, Mark Hutchinson, Tony and Katie Hiller, Rod Hobson, Dave Knowles, Andrew O'Grady, Mike Powell, Peter Robertson, Jo Sumner, Gerry Swan and Mike Swan. We also thank the herpetology research group at Museums Victoria for their ideas and Dermot Henry for his strong support of this book from its beginnings.

Mark Hutchinson reviewed parts of our manuscript and offered some extremely sound advice. We thank him for sharing his immense herpetological knowledge.

Jane Melville dedicates this book to the memory of her dad, who passed on his enjoyment of nature and the outdoors, he would have been very proud. Also her mum for love and support over the years. Her children Eva and Thomas for their never-ending energy and Vern and Lois, loving grandparents, for making many of these fieldtrips possible through kind hours of childcare in strange locations. Finally, for her wonderful husband, Andrew O'Grady, whose expertise in fieldwork, endless support and sound advice has made this book possible. Thankyou.

Steve K. Wilson would like to thank his long suffering wife, Marilyn Parker, who continues to indulge his obsessive passion, and even shared her honeymoon with a quest to find a rare Greek viper. His parents Joy and Ken Wilson gave him free rein and limitless encouragement to pursue his dreams.

# BIOGRAPHIES

**Jane Melville** has always been fascinated by reptiles, spending much of her childhood in Tasmania exploring bushland in search of lizards. She went on to undertake a PhD at the University of Tasmania on the evolution of snow skinks, graduating in 1998, then turning her focus to dragon lizards. In 2002 she started at Museums Victoria, where she is the Senior Curator of Herpetology. Her research career has taken her to almost every continent and every corner of Australia studying the evolution, taxonomy and conservation of dragon lizards. She has now named more than 15 new species. Her expertise on Australian dragon lizards is internationally recognised, with more than 70 research papers, she provides advice on taxonomic and conservation issues, and is a past President of the Australian Society of Herpetologists. She is dedicated to training young researchers and communicating her knowledge of lizards to a wide audience.

**Steve K. Wilson's** childhood passion for wildlife started with plastic dinosaurs, catching cicadas and patrolling the local suburban creek for skinks and frogs. He now works as an independent fauna consultant, and is employed by the Queensland Museum as an Information Officer, where he answers public inquiries and identifies a diverse range of natural history specimens. He is the author and co-author of nine books on herpetology, covering regional, state and Australian field guides, lizard natural history and a childrens' book. His images and articles are published widely in magazines in Australia and overseas. An award-winning photographer, he was nominated by Australian Geographic magazine as their 'Photographer of the Year' for 2000. Steve's interest in natural history and photography has led to extensive travel across Australia and to many of the worlds' unique biodiversity hotspots.